22-95

Feminist Futures?

Performance Interventions

Series Editors: **Elaine Aston**, University of Lancaster, and **Bryan Reynolds**, University of California, Irvine

Performance Interventions is a series of monographs and essay collections on theatre, performance, and visual culture that share an underlying commitment to the radical and political potential of the arts in our contemporary moment, or give consideration to performance and to visual culture from the past deemed crucial to a social and political present. *Performance Interventions* moves transversally across artistic and ideological boundaries to publish work that promotes dialogue between practitioners and academics, and interactions between performance communities, educational institutions, and academic disciplines.

Titles include:

Alan Ackerman and Martin Puchner (*editors*)
AGAINST THEATRE
Creative Destructions on the Modernist Stage

Elaine Aston and Geraldine Harris (*editors*)
FEMINIST FUTURES?
Theatre, Performance, Theory

Lynette Goddard
STAGING BLACK FEMINISMS
Identity, Politics, Performance

Leslie Hill and Helen Paris (*editors*)
PERFORMANCE AND PLACE

Melissa Sihra (*editor*)
WOMEN IN IRISH DRAMA
A Century of Authorship and Representation

Forthcoming titles:

Amelia M. Kritzer
POLITICAL THEATRE IN POST-THATCHER BRITAIN

Performance Interventions
Series Standing Order ISBN 1–4039–4443–1 Hardback 1–4039–4444–X Paperback
(*outside North America only*)

You can receive future titles in this series as they are published by placing a standing order. Please contact your bookseller or, in case of difficulty, write to us at the address below with your name and address, the title of the series and the ISBN quoted above.

Customer Services Department, Macmillan Distribution Ltd, Houndmills, Basingstoke, Hampshire RG21 6XS, England

---

Feminist Futures?

Theatre, Performance, Theory

Edited by

Elaine Aston and Geraldine Harris

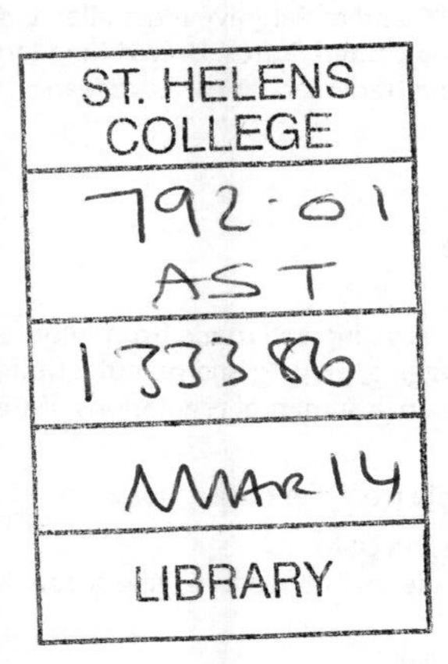

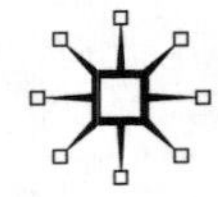

First published 2006
This paperback edition first published 2007 by
PALGRAVE MACMILLAN
Houndmills, Basingstoke, Hampshire RG21 6XS and
175 Fifth Avenue, New York, N.Y. 10010
Companies and representatives throughout the world

PALGRAVE MACMILLAN is the global academic imprint of the Palgrave Macmillan division of St. Martin's Press, LLC and of Palgrave Macmillan Ltd. Macmillan® is a registered trademark in the United States, United Kingdom and other countries. Palgrave is a registered trademark in the European Union and other countries.

ISBN-13: 978–1–4039–4532–7 hardback
ISBN-10: 1–4039–4532–2 hardback
ISBN-13: 978–1–4039–4533–4 paperback
ISBN-10: 1–4039–4533–0 paperback

This book is printed on paper suitable for recycling and made from fully managed and sustained forest sources. Logging, pulping and manufacturing processes are expected to conform to the environmental regulations of the country of origin.

A catalogue record for this book is available from the British Library.

Library of Congress Cataloging-in-Publication Data
Feminist futures? : theatre, performance, theory / edited by Elaine Aston and Geraldine Harris.
p. cm. — (Performance interventions)
Includes bibliographical references and index.
ISBN 1–4039–4532–2 (cloth) 1–4039–4533–0 (pbk)
1. Women in the theater. 2. Feminist theater. 3. Feminist drama, English—History and criticism. I. Aston, Elaine. II. Harris, Geraldine (Geraldine Mary) III. Series.
PN1590.W64F44 2006
792.082—dc22 2005055268

10 9 8 7 6 5 4 3 2 1
16 15 14 13 12 11 10 09 08 07

Printed and bound in Great Britain by
Antony Rowe Ltd, Chippenham and Eastbourne

To feminist futures

Contents

Acknowledgements

This collection would not have been possible without the funding assistance of the Arts and Humanities Research Council (AHRC) and the inspiration from artists and participants in our Women's Writing for Performance project. A big thank you to our administrator Susie Wood for her assistance and her patience.

A special note of thanks to all the scholars and practitioners who got involved in and contributed to this project – especially artists Lenora Champagne, Clarinda Mac Low, Ruth Margraff and Fiona Templeton, who turned out on a freezing-cold Sunday afternoon in New York to talk 'feminist futures'.

Also thanks to all of our colleagues in Theatre Studies and the Nuffield Theatre and to Mike Bowen of the Lancaster TV unit for his patience and good humour in helping with audio technology.

Finally: for Gerry, as ever, thanks to Colin Knapp – (for the sentimental, which excludes Colin himself – see Jacques Brel, *La chanson des vieux amants*). For Elaine – a thank you to my mother June for her continued love and support, and to Maggie and Daniel for surviving burnt dinners and office days, but without whom nothing would make any sense at all.

Notes on Contributors

Elaine Aston is Professor of Contemporary Performance at Lancaster University where she researches and teaches feminist theory, theatre and performance. Her authored studies include *Sarah Bernhardt: A French Actress on the English Stage* (Berg, 1989), *Theatre as Sign-System* with George Savona (Routledge, 1991), *An Introduction to Feminism and Theatre* (Routledge, 1995), *Caryl Churchill* (Northcote, 1997; 2001), *Feminist Theatre Practice* (Routledge, 1999) and *Feminist Views on the English Stage* (Cambridge, 2003). With Gerry Harris she is currently leading a three-year AHRC-funded project, 'Women's Writing for Performance'.

Sue-Ellen Case is Professor and Chair of Critical Studies in Theatre at UCLA. Her books include *Feminism and Theatre* (Routledge,1988), *The Domain-Matrix: Performing Lesbian at the End of Print Culture* (Indiana University Press, 1996) and *From Alchemy to Avatar: Theatre, Science and the Virtual* (2006). She has edited several critical anthologies and play collections including *Split Britches: Lesbian Practice/Feminist Performance* (Routledge, 1996), *Performing Feminisms* (Johns Hopkins University Press, 1990) and a series of critical collections entitled 'Unnatural Acts: Theorizing the Performative' (Indiana University Press). She is also the author of numerous articles on German theatre, feminist performance and lesbian theory.

Anna Furse is an award-winning director and writer of over 50 text-driven and devised works. A classically trained dancer influenced by study with Peter Brook, Grotowski and new dance forms, she has developed her own training methodology that creates theatre from the body outwards. Artistic directorships include the feminist company Blood Group, Paines Plough and her new company Athletes of the Heart. Her published plays include *Augustine (Big Hysteria)* (Harwood Academic, 1997) and 'Gorgeous' (Aurora Metro, 2003). Her productions have toured Europe, Asia and the United States. She has taught for 27 years in the United Kingdom and internationally. She is currently a full-time lecturer in the Department of Drama at Goldsmiths College, University of London, where she directs the MA in Performance.

Geraldine (Gerry) Harris is Professor of Theatre Studies at the University of Lancaster. Published work includes articles and book chapters on topics ranging from female performers in nineteenth-century French popular theatre to issues for contemporary performance practice. She has written two monographs, *Staging Femininities: Performativity and Performance* (Manchester, 1999) and *Beyond Representation: The Politics and Aesthetics of Identity in Television Drama* (Manchester, 2006). Since the 1980s she has also engaged with practice as a writer and devisor/director. With Elaine Aston she is currently working on a three-year AHRC-funded project, 'Women's Writing for Performance.

Dee Heddon is a senior lecturer in Theatre Studies at the University of Glasgow. Her research on autobiographical performance has appeared in various journals including *Studies in Theatre Production, Research in Drama Education, Performance Research, New Theatre Quarterly* and in anthologies including *Auto/biography and Identity* (Manchester, 2004) and *Modern Confessional Writing* (Routledge, 2005). Dee's autobiographical performance practice includes *Forward to Sender* (2002), *Following in the Footsteps* (2003) and *One Square Foot* (2004). She is also co-author of *Devising Performance: A Critical History* (Palgrave Macmillan, 2005).

Leslie Hill and **Helen Paris** are writers, performers, filmmakers and co-directors of Curious (www.placelessness.com). Curious was formed in 1996 and has since toured over 30 performance and film and video projects in Europe, North America, Australia, Brazil, China and India. Their solo performances *Smoking Gun* (Hill) and *Family Hold Back* (Paris) were recently presented by the Sydney Opera House. Their new film, *Red Lantern House*, was shot on location in Shanghai with support from the British Council Artist Links programme. Curious recently received an artist residency in Paris at Couvent des Récollets, supported by the French Ministry of Cultural Affairs and the Mairie de Paris. They have published a joint-authored book *Guerilla Performance: How to Make a Living as an Artist* (Continuum, 2004) and have co-edited a critical anthology *Performance and Place* (Palgrave Macmillan, 2006). Leslie Hill is a NESTA Dream Time Fellow.

Aoife Monks is a lecturer in theatre at the School of English and Humanities, Birkbeck College, University of London. She previously taught at the Department of Film, Theatre and Television at the University of Reading. Aoife's research interests centre on the actor's body in performance and she is currently working on a monograph analysing the use of costume and make-up in contemporary theatre practice. Other research interests include the work of the Wooster Group and the contemporary performances of Irishness. She has published articles in

Modern Drama, *Australasian Drama Studies* and *New England Theatre Journal*. She has also worked as an intern with the Wooster Group in New York.

Meenakshi Ponnuswami is Associate Professor of English at Bucknell University, where she teaches dramatic literature and theatre history. Her scholarly work has focused on representations of history in contemporary British theatre. She has written on Howard Brenton, Caryl Churchill and other contemporary British dramatists. These publications include a chapter in The *Cambridge Companion to Modern British Women Playwrights* (Cambridge University Press, 2000) and she has edited the critical collection *Contemporary British Theatre: The New Left and After* (Routledge, 2002). She is currently studying the plays of Winsome Pinnock.

Janelle Reinelt is Associate Dean of Graduate Studies and Professor of Drama at University of California, Irvine. She is the current President of the International Federation of Theatre Research and co-editor with Brian Singleton of the book series 'International Performance and Culture' (Palgrave Macmillan). Her books include The *Cambridge Companion to Modern British Women Playwrights* with Elaine Aston (Cambridge University Press, 2000), *After Brecht: British Epic Theatre* (University of Michigan Press, 1994), *Crucibles of Crisis* (University of Michigan Press, 1996) and *Critical Theory and Performance* co-edited with Joseph Roach (University of Michigan Press, 1992; 2006). She is a former editor of *Theatre Journal*.

Lena Šimić is a performance artist and is currently undertaking a practice-based PhD in the Institute for Contemporary Arts at Lancaster University. Lena was born in Croatia and trained in theatre directing and acting at the Academy of Music and Dramatic Arts in Bratislava, Slovakia, and the London Academy of Performing Arts. Recent performance projects include *Medea/Mothers' Clothes, Magdalena Makeup* and *Joan Trial*; a series of interventions into female archetypal figures. Lena has collaborated with the Bluecoat Arts Centre in Liverpool, the Nuffield Theatre in Lancaster and the Art Workshop Lazareti in Dubrovnik. She has toured her performances nationally and internationally.

SuAndi is of Nigerian and British heritage and has been a Performance Poet since 1985. Her collections of poems include *Style* (Purple Heather & Pankhurst Press, 1990), *Nearly Forty* (Spike Books, 1994), *There Will Be No Tears* (Pankhurst Press, 1996) and *I Love the Blackness of my People* (Pankhurst Press, 2003). She has also created work for the Live Art stage and her last ICA commission, *The Story of M*, received critical acclaim in the United Kingdom and North America. In addition she has written two librettos, *The Calling* (BBC Philharmonic, 2005) and *Mary Seacole Opera* (2000), which toured Britain after a West End opening. Her work has been recognized with a NESTA Fellowship in 2005, The Big Issue Community Diploma 2003, The Windrush Inspirational Award 2003 and the OBE in the Queen's 1999 New Year Honours List, following her Winston Churchill Fellowship in 1996. Since 1985, SuAndi has been the freelance Cultural Director of Black Arts Alliance. BAA is the United Kingdom's largest and longest surviving network of Black artists.

Joanne Tompkins teaches drama at the University of Queensland, Australia. Her books on post-colonial, intercultural, and multi-cultural theatre and theory include *Post-Colonial Theatre* with Helen Gilbert (Routledge, 1996) and *Women's Intercultural Performance* with Julie Holledge (Routledge, 2000). She is currently completing *Unsettling Space* (Palgrave Macmillan, 2006). In addition to theoretical research on spatiality, she generates virtual-reality models of actual theatres for both research and for practical applications. She has completed a five-year term as co-editor of *Modern Drama*, which also included co-editing *Modern Drama: Defining the Field* with Ric Knowles and W. B. Worthen (University of Toronto Press, 2003).

1

Feminist Futures and the Possibilities of 'We'?

Elaine Aston and Geraldine (Gerry) Harris

In the concluding remarks of the introduction to her highly influential book *Feminism and Theatre*, Sue Ellen Case observed:

> I think it [*Feminism and Theatre*] suggests the radical way in which feminism has affected all aspects of theatre, changing theatre history and becoming a major element in twentieth-century theatre practice. The feminist critic or practitioner need no longer adopt a polemic posture in this art, but can rely on the established feminist tradition in the theatre, with its growing number of practitioners and adherents.[1]

Case explains that she was writing *Feminism and Theatre* in 1985, 'when the feminist movement was only twenty years old and feminist theatre practice younger still' (p.4).

Another 20 years on, *Feminist Futures*? is a far less confident enterprise, even though in Case's genealogy feminism and feminist theatre practice have barely achieved middle-age. Comprising a collection of essays by practitioners and scholars, including Case herself, it stages a (sometimes polemic) debate on *if* and in what ways feminism may still be an element (major or otherwise) of theatre and performance practice of the twenty-first century. The question mark in the title of our volume is crucial. This question mark poses the future of feminism and the relation between feminism and theatre and performance as a question and as being *in* question. In begging the question as to whether or not feminism may still obtain as a mobilizing force shaping political, social and artistic futures, it invites an interrogation of the present, which in turn demands some reflection on the legacies of the 'established' feminist and feminist-theatre traditions of the past.

Our line of questioning is presented at a time when Western feminism has no high-profile political movement, and when debates about feminism, in both the public and academic spheres, circulate in a climate of 'postfeminism'. As is evident from the proceedings at a number of recent conferences focusing primarily or partly on postfeminism (such as those held in Britain at Exeter University (2003), the University of East Anglia and the University College of Northampton (both in 2004), and at Cambridge in (2005)), there is significant debate over the genesis and definition of this term. In the public sphere, as Susan Faludi notes in *Backlash*, 'postfeminist sentiments' were already in circulation in the 1920s press, just after women first got the vote in North America and Britain, re-emerging with increasing strength in the media of the1970s and the 1980s.[2] In the academic sphere, starting mainly in media studies in the late 1980s, the term postfeminism is often conflated with 'third wave' or postmodern feminism,[3] even though this is not a term employed by the most influential 'postmodern' feminist thinkers of this period.

In public and academic spheres, postfeminism is usually defined in opposition to 'second wave' feminism. This 'oppositional' stance is frequently defined in 'generational' terms. According to critics such as Natasha Walter, advocate of the 'new feminism', a younger feminist generation needs to reject the 'rigid ideology' of a seventies-style feminism that, in Walter's view (and others of a similar persuasion), 'led feminism to a dead end'.[4] While Walter and colleagues argue for a revitalized form of 'new feminism', there is also a growing (popular and academic) consensus around the idea of 'postfeminism': that this prefix 'post', points to feminism that, for good or ill, has done its job and is now redundant, over, finished, no longer required. Many university women's studies departments are shutting or are under threat, and within the media, feminism has been constantly demonized (feminist-as-man-hater), or reactivated through a conservative, rather than radical political discourse (feminism's transformation from bra-burning to Wonderbra advertisements, for example). It would appear, then, that younger generations of women do not 'need' and/or prefer to disown and/or cannot identify with feminism and therefore, presumably, with 'the established feminist tradition in the theatre' (see Leslie Hill and Helen Paris, Chapter 4, p.57).

Bearing all of this in mind, in setting out to commission contributions for this volume, our brief as editors was to try, as far as possible, to attract contributions from different generations of women scholars and practitioners. Equally, rather than invite essays that would make the case *for* feminism and its relation to theatre, we attempted to encourage debate, reflections on, dialogue with feminism, theatre, performance and futures, allowing for the possibility of taking up a range of views and positions. That said, as is undoubtedly already evident, as editors, we have, to use Sarah Ahmed's term, 'feminist attachments', as do all of the contributors in this volume (albeit in very different ways).[5] Briefly stated, in our view, this means an enduring 'attachment' to certain ideals as remaining important and necessary to improving the social and cultural welfare of 'women's' lives, never more so perhaps, in a period where it is no longer possible to overlook the connections between the local and the 'global'. As such, we take issue with embracing the term *post*feminism which, as Janelle Reinelt argues in her essay, risks the 'performatively defeatist' suggestion of 'giv[ing] up on the project of feminism' (Chapter 2, p.17). There is plenty of concrete evidence to suggest that even the most privileged do not as yet inhabit a world in which violence, injustices and inequalities are no longer carried out in the name of identity categories.

However, coming as we do as editors from a background in the theatre academy that 'grew up' with feminism, surfing *both* its second and third 'waves', it is important to state at the outset that this volume is not a nostalgic lament for a feminist past, but an engagement with current directions – even if these sometimes seem to be inadequate, obscure, difficult, hard to find or determine. Feminism has always operated self-reflexively: as an evolving 'body' of political ideas and impulses. Feminism is not, therefore, something we propose as us needing to 'get back to': it is not our nostalgic longing for a feminism we knew, experienced and lived before. Essential to any sort of feminist politics has always been the idea that the 'future' is a question, is *in* question: is not necessarily determined by the past or the present. Rather, then, our desire is for a feminism to move forward with, or to keep us moving forward. As Ahmed argues, '[t]o stay open to feminism is both to critique the world, which we face in the present, and to encounter the objects of feminism anew, as that hope for the "not yet", in the here and now' (*Cultural Politics of Emotion*, p.187).

Contemporary feminist theatre-making has constantly represented and presented both utopic and dystopic social realities in the interests of making the 'not yet' visible through the theatrical imagination (see Sue Ellen Case, Chapter 7, and Joanne Tompkins, Chapter 12). Crucially, such feminist theatrical imaging and imagining of the 'not yet' has had a strong and enduring attachment to collaborative or collective models of theatrical staging and aesthetics. The democratization of theatre labour and aesthetics is allied to the realization of the 'not yet' of a feminist politics concerned with alternative (more democratic) futures of all, rather than one, or some, women. In brief, the suggestion is that it is only by working together and by painstaking negotiation across sites of difference, that the 'not yet' may come to figure as a possibility. It is arguably harder to do this kind of collaborative feminist work in today's political and cultural climate that is hostile to any kind of left-wing ideology or politics. Yet, Ahmed maintains:

> The openness that gathers in the struggle against 'what is' involves the coming together of different bodies in this present time. It is here that the feminist 'we' becomes affective. For the opening up of that which is possible does not take place in time, in that loop between present and future. The opening up *also takes time.* The time of opening is the time of collecting together.
>
> (*Cultural Politics of Emotion*, p.188)

'Collecting together' the different voices in *Feminist Futures?* is our attempt at 'opening up' the different sites of feminism, theory, theatre and performance in this present time to interrogate what has been and what is, and to consider the possibilities of what might be. Given that, as Ahmed notes, such opening up *takes time,* we cannot possibly address these questions in relation to the whole field, or rather fields, of feminism, theatre and performance in a single volume, let alone in this introduction. As a way of 'opening up' this collection, we necessarily have to restrict the possibilities of framing and of introducing. The introductory focus we have chosen, therefore, is to look back at some of the 'futures' proposed by feminism in theatre, performance and theory, and at the different (mobile) generations of feminism and feminist theatre that have (to borrow from Ahmed) taken 'the risk of inhabiting its [feminism's] name' (*Cultural Politics of Emotion*, p.188).

Embracing 'we'

Hope that, in the future, 'what is' might be radically transformed, revolutionized in the interests of women's 'liberation', characterized a 1970s style of Western feminism that argued its 'demands' for women's lives to be released from oppressions brought about by the dominant forces of patriarchy and capitalism. Feminism brought women together, united them, gave them a 'group' identity. Consciousness-raising groups allowed women to meet together, providing a forum for 'raising' the personal 'I' into a political, collective 'we' (see Dee Heddon, Chapter 9, p.131). As individual women joined other women to 'name' their particular, personal problems and grievances, they initiated a collective, political naming of the injustices, the inequalities against their sex. In identifying as 'women', they identified 'men' as the enemy. Feminist theorist Hélène Cixous, whose writings influenced a generation of European theatre makers, explained how '"[w]e" struggle[d] together', fighting 'discrimination, the fundamental unconscious masculine racism', an 'experience that launched the front line of feminist struggle in the United States and in France'.[6]

The 'front line of the feminist struggle' found its way into all kinds of cultural activity. In theatre specifically, it encouraged women to form their own, often sex-segregated, companies such as The Women's Theatre Group and Monstrous Regiment in the United Kingdom, for making work primarily by and for women. In this feminist theatre practice men were dramatized as the 'enemy', while women moved 'centre stage'. All of this was frequently accompanied by a rejection of realist forms as expressive of a masculine, patriarchal aesthetic. History was re-staged through a feminist lens and mostly a Brechtian-feminist aesthetic to demonstrate past oppressions and to give dramatic expression to a feminist plea and urgency not to revisit, not to allow the oppressions of the past a place in future histories. At the same time, under the influence of theorist-practitioners such as Cixous, the search began for a 'new' theatre language: a feminist poetics which would challenge the theatrical apparatus, its systems of representation and narrativization that positioned women as Objects and Others. Symbolic framings of women as Woman were contested as feminist theatre practitioners refused to be objectified, gazed upon – refused to be the 'other' in some(male)body else's story. In the fields of live and performance art there was a dynamic explosion of feminist work, exploring the potential of the body as a scene of, and screen for, 'writing' histories of gender oppressions and for the possibilities of 'speaking' the feminine: marking present tense (lived) oppressions with the desires, the longings for lives lived differently: beyond the unequal social realities conditioning women's lives.

While the emotional mix of angry pasts and brighter, more hopeful futures determined the motivations and actions of characters in the feminist drama, or scripted the 'bodies' of feminist performance artists, it was also designed to 'work on', to persuade spectators. In parallel with the consciousness-raising group model, feminist theatre proposed a collective naming and sharing of problems and oppressions, one that might draw the individual into collective action: the personal (individual) hooked up to political (group) agency. It was a model of feminist spectatorship that assumed a 'closed' circuit of feminism between performers and audience. Moreover, to be persuaded by feminism, to be of a feminist persuasion, did not end in the theatre: 'newly born' and committed feminists felt their lives, emotions, thinking and politics touched by, moved by, the imagined worlds of feminist theatre and performance.

The violence of 'we'

While there were a good number of women who took the 'risk of inhabiting its [feminism's] name' during the 1970s, equally there were many who were careful about their 'attachment' to feminism, as evidenced in the growing emergence and adoption of the phrase 'I'm not a feminist but...' – commonly employed by a large number, even perhaps the majority of Western women, including many working in theatre. Alternatively, some female practitioners, especially those attempting to gain a '*toe*hold' in mainstream theatre, employed the variant 'I'm a feminist but...' when discussing their practice. This disclaimer was motivated by the desire to avoid a categorization that might influence or predetermine the reception and perception of their work, whether within the still patriarchal institution of theatre or due to the 'burden of representation' that might be placed upon it by other feminists (see Aoife Monks, Chapter 6).

More problematic than this, however, was the way in which second-wave attachments to feminism meant that, despite the best of intentions, the collective 'we' generated by the movement tended to be exclusive rather than inclusive. In brief, in assuming the 'essential' priority of (inter)sexual difference, feminism generated a 'we' that failed to take account of how it might be simultaneously inscribed through discourses of class (middle), sexuality (hetero) and above all 'race' (white).

Throughout the 1980s, the exclusivity of this feminist 'we' (particularly given the gradual rise of an individualist (right-wing) style of feminism, that we would argue remains unabated) was variously contested by those whom it failed to include. In Britain the dominance of socialist-feminism in both theory and in feminist theatre practice kept issues of class on the agenda, at least until the early 1990s. In terms of sexuality, the exclusivity of the feminist 'we' was challenged early on by the emergence of influential lesbian feminist theorists such as Monique Wittig and Teresa De Lauretis, to name but two. In the sphere of theatre it was challenged by the inauguration of companies like Gay Sweatshop (1975), with its strong links to the Women's Theatre Group (1974), and by venues such as London's Drill Hall (1978) and the celebrated WOW café in New York (1980). Both these venues were crucial spaces for an increasingly visible and radical generation of lesbian-feminist artists and theorists whose radicalism again extended to experiments with form dedicated to resisting the heterosexual imperative traditionally inscribed within the structures of realism and naturalism.

However, much second-wave feminism seldom engaged with the problematics of 'race' except through class and gender. In her essay on 'the violence of "we"', feminist theorist and theatre critic Elin Diamond argued and gave apposite examples of the ways in which a white Western feminist perspective repeatedly tended to elide the representation of black women.[7] As Diamond observed, what was 'revealing' was the way in which feminists who 'thought they were producing new sites of "we" – a we that empowered women', had 'in fact excluded many women' ('The Violence of "We"', p.393).

In 1982, as a response to the violence of this feminist 'we', Hazel V. Carby demanded 'White Woman Listen' to her articulation of 'black' feminism and of the boundaries and limits of sisterhood.[8] Similarly, writing in 1984, bell hooks challenged the liberal, bourgeois base to a predominantly white feminist movement that, in her view, had failed to attend to matters of race and class privilege.[9] Interviewed in *Borderlands/ La Frontera* (1987) about her co-editorship (with Cherie Moraga) of the seminal collection of writings by radical women of colour, *This Bridge Called My Back* (1983), the late Gloria Anzaldua described her motivation behind the collection as a response to the experience of being spoken for (rather than listened to) by white feminists; of an eradication of cultural and social differences by women who 'never left their whiteness at home'.[10] While 'great' and 'supportive' on the one hand, white feminist women, on the other, 'blacked out and blinded out... oppressions': 'they had an idea of what feminism was and they wanted to apply their notion of feminism across all cultures' (*Borderlands*, p.231).

In theatre, the 'blacked out' views of a dominant white feminism prompted 'othered' women to make, write and perform works exploring identities informed not only by gender, but by the complex *intersections* of race, class and sexuality. This was reflected in the formation of companies such as Spider Women in the United States (1975) and Theatre of Black Women (1982) in the United Kingdom, and in works by practitioners such as Ntozke Shange and Robbie Macauley in the United States, or UK dramatists Jacqueline Rudet, Jackie Kay and other poets/performers, such as SuAndi who contributes to this collection. Again this work often (although not always) eschewed realist forms as expressive of a white, Western, bourgeois subjectivity, in favour of strategies that reflected the specificity of black women's cultural experience and legacies, including those arising from inhabiting 'multiple and interrelated identities'.[11] By the mid-1990s this 'multiple' position was often expressed in terms of anti-essentialist postcolonial theory and by means of a 'hybrid' diaspora aesthetics (see Meenakshi Ponnuswami, Chapter 3). It is notable that, despite their criticism of the ways the socially and culturally dominant feminist 'we' elides differences between women, many of these 'black' and 'lesbian' and queer theorists and practitioners continued to identify as feminists.

'We' and the category of 'women'

Under the influence of these voices, ('white') Western feminism started to produce its own self-reflexive critiques of the way its 'we' failed to take account of differences within the category 'women'. Within the academic sphere works like Donna Haraway's 'A Manifesto for Cyborgs', originally published in 1985,[12] drew on poststructuralist and postmodern theory to deconstruct feminist essentialism and the subsequent privileging of the white Western, middle-class female subject. However, Haraway also insisted on the continuing importance of the legacy of materialist feminism in a world undergoing the processes of 'globalization', speeded by new technologies and dominated by militaristic Western capitalism. In order to represent the (future) possibilities for a posthuman, anti-racist, anti-colonialist, queer feminist politics, Haraway constructs a 'science fiction' of cyborgs. This manifesto is written from a dual perspective: at once apocalyptic and utopian, it explores the material inequalities and very real dangers of the present even as it imagines a brighter future, the realization of which depends upon feminists actively engaging with, and transforming the use, place and meaning of, emergent technologies. As such, it became a key text for academics and practitioners concerned with politically engaged multi-media performance (see Anna Furse on 'The Theatre of Technoscience', Chapter 10, pp.163–6).

In 1990 Judith Butler published *Gender Trouble, Feminism and the Subversion of Identity*, closely followed in 1993 by *Bodies that Matter: On the Discursive Limits of 'Sex'*. Like Haraway, Butler can be said to work from a 'dual perspective' that is at once future utopian and deeply pragmatic in the here and now. As she points out, taking up the normative sexed/gendered positions remains (at present) one of the conditions that 'qualifies a body for life within the domain of cultural intelligibility'.[13] At the same time, Butler analyses these positions as 'regulatory fictions', in ways which allow for their destabilization and 'denaturalization' through a process of 'unfaithful' or 'subversive repetition' that foregrounds the undecidable play of multiple and plural differences within identity.

Yet even as Butler offered a deconstruction not just of 'gender' but also of 'sex' and therefore of the foundations of the category 'women', like Haraway she remains committed to the idea that, however flawed in the past or the present, feminism (which itself depends on that category) still has an important role to play in creating a more just and democratic future. Butler explains this contradiction through reference to postcolonialist theorist Gayatri Spivak, in terms of a 'contingent essentialism' that recognizes the necessity 'to invoke the category and, hence, provisionally to institute an identity' so as 'to open the category as a site of permanent political contest' (*Bodies that Matter*, pp.221–2).

By the 1990s, then, these contributions from queer, anti-racist feminists to resist, counter or deconstruct the violence of the white Western feminist 'we' did sometimes seem to signal the possibility of utopian futures beyond or 'post' the politics of identity. Moreover, the focus in many of these discourses on the cultural and the aesthetic, on strategies of mimicry, on the foregrounding of subjectivity and identity as multiple, artificial and constructed, and on the possibilities of 'performative' transformation, all seemed to point to theatre and performance as an ideal site for the exploration and embodiment of these ideas: as a space not just for imagining but actively producing this future.[14]

'We' no more

Somehow these imaginings of a utopian future became translated into a fantasy of the present, in which the deconstruction of the Enlightenment humanist subject within the academic sphere and in certain types of cultural practice, such as theatre and performance, means that there is 'no longer any hierarchy for the disenfranchised to undermine'.[15] It appeared, to borrow a phrase from Jane Flax, as though the matter of differences had been 'dealt with',[16] their proliferation so effectively noted, embraced and celebrated that the category 'women' has been conclusively shown to be a hegemonic discursive fiction. This spells the 'end' of feminism and since *any* articulation of a feminist 'we', however contingent, may be characterized as potentially exclusive, as a discourse feminism might even be invoked as a *source* of oppression for the multiple and plural subjects, previously known collectively as 'women'.[17]

Yet in many ways, as the controversial performance artist Karen Finley argued passionately and angrily about US capitalism in her mid-1980s solo show *The Constant State of Desire*, it seems (and feels) as though 'nothing happened'.[18] This is perhaps not surprising since, as Butler points out, while the sex /gender positions may indeed be historical 'regulatory fictions', they nonetheless continue to have concrete social and material effects (*Bodies that Matter*, p.10).

There is no question that *some* things have improved to *some* extent for *some* women, although, strangely enough, the vast majority of these appear to be white, Western and middle class. Faced with the unshakeable (patriarchal) dominance of Western capitalist, and increasingly trans-capitalist's organizational structures, values and 'investments' in materialistically driven local and global economies, feminism has not revolutionized or effectively combated the majority of the injustices, cruelties and inequalities that it brought to public attention in the 1970s. Nor has it yet challenged the 'new' ones that have arisen as a result of these processes. Yet it is important to remember, as Susan Bordo points outs, feminist theory – even where it is the work of white, middle-class, heterosexual women, 'is not located at the *center* of cultural power' (emphasis in the original).[19] As such, we would argue against the dangers of postfeminist mythologies that, as Susan Faludi and other feminists argue, would have us believe in a here and now future in which all oppressions have been deconstructed.

In fact, we have concerns about the appropriation and inoculation of feminism within the (still) patriarchal set up of the academy, which, under government pressure, seems to be becoming more and more conservative. Repeatedly, for instance, we have heard Haraway's manifesto appropriated in ways that ignore or underplay its commitment to feminism, and reduce its dual perspective to a single vision that declares a cyborgian utopia NOW. In performance studies, at its most extreme this produces a techno-romantic formalism that assumes that if a work is produced through or employs 'new' technologies, it is somehow automatically and inherently politically 'radical'. There is little recognition of Haraway's materialist critique of the role of technology in the 'informatics of domination' and of the conditions of its production by a new global underclass. Nor is there recognition that, as Haraway's discourse makes clear, robbed of its oppositional politics, the 'cyborg' exactly fits the requirements of aggressive global, militaristic capitalism.

As Sara Ahmed indicates in *Strange Encounters*, the same processes of inoculation and appropriation have also occurred in postcolonial theory, where the prefix 'post' is sometimes assumed to indicate an *overcoming* or transcending of the historical legacy of colonialism, including the construction of the Enlightenment subject.[20] This then becomes grounds for the celebration of 'new' forms of global(ized), 'nomadic' subjectivity. In fact, as Ahmed asserts, 'what is at stake here is a certain kind of Western subject, the subject of and in [postmodern] theory who is free to move' – at the *expense* of others (*Strange Encounters*, p.83). Morever, Ahmed points out that within this theory, 'race' and 'class' are still more often employed as 'figures for difference', rather than as a 'constitutive and positive term of analysis' (*Strange Encounters*, p.41).

Similarly, Butler's thesis in *Bodies That Matter*, relating to the performativity of the sex/gender positions, appears to have led to a number of assumptions, not least that the term performative actually signifies as 'inherently deconstructive' and that strategies of irony, parody and pastiche are always 'resistant', as if these modes cannot and have not been employed, in the past and present, to serve reactionary political agendas.

These assumptions are especially misleading in the realm of theatre and performance, where the term 'performative' is now often used as interchangeable with the terms performance and/or theatrical. This potentially fosters the illusion that all 'repetitions' of gender roles in these sites are equally 'resistant', and alongside this that the imagining or enactment of transformations of the sex/gender system in this sphere is enough in itself to (performatively – as in instantly) effect a transformation in the world of the social. As queer theorist Alan Sinfield, discussing Homi K. Bhahba alongside Judith Butler, argues 'We have supposed too readily that to demonstrate the indeterminacy and weakness in a dominant construct is to demonstrate its weakness and vulnerability to subversion ... But in actuality as Marx tells us, capitalism thrives on instability.' He continues: 'It is easier than we once imagined to dislocate language and ideology and harder to get such dislocations to make a practical difference.'[21]

Contributions to the possibilities of 'we'?

In the analysis of what might be designated as the utopic (post)feminist moment, fixated on imagining the future while (we would argue) ignoring the material exigencies of the present, we are also perhaps tending too far towards the dystopic. As many of the contributions in this volume attest, in recent years numerous women practitioners in different countries from different generations have produced vital and challenging works of theatre and performance that address a wide range of pressing social and political problems for women, touching the local and the global. These contributions also attest to the continued existence of theorists, including those working in the field of theatre and performance, who continue the process of self-reflexive critique that has always been fundamental to feminism moving forward.

Nevertheless, for all the reasons cited above, this seems a crucial time to ask if the struggle for an affective and just feminist 'we' is indeed no longer politically necessary, useful, meaningful or desirable? This seems especially pressing in the light of recent whispers of the beginnings of something called the 'new universalism' within the field of theatre and performance criticism. This stance appears to surface as a response to the failure of some 'postmodern' theory to fully address the material and political affects of globalization, its failure to offer a way of negotiating the differences, conflicts, antagonisms and injustices that have arisen – except in terms of liberal relativism. We (as editors) are definitely not advocating a return to 'universalism' – neither in a 'new' guise nor an old one. Yet as we note in our collaborative chapter with Lena Šimić, we do have concerns over the effects of what Sedgwick describes as 'anti-essentialist theoretical hygiene'. This often appears to have resulted in a 'new' individualism, which, paradoxically, goes along with 'universalism': the undermining of a sense of 'we' as a contingent, collective *political* position that is simultaneously an undermining of a sense of agency – an ability to act on and change the world.

What also seems to us to have 'got lost' in the postmodern, postfeminist utopian moment is the understanding that differences (ethnic, sexual, class, sexuality, age, religion, national, etc.) cannot be 'dealt with' instantly in a single performative gesture or through a series of 'stylized acts'. Neither can they be 'dealt with' by simply listing them, embracing them, celebrating them or remarking their proliferation. As is evident in both the micro-personal and macro-political spheres, encountering, engaging with, negotiating and living with and alongside differences, is extraordinarily demanding. It is also risky: there is the risk of failure, of antagonism, of misunderstanding, of pain; a risk that the sense of 'self' (however this is understood) might have to move, might *be* moved.

In the conclusion to her book *Strange Encounters*, when Ahmed models a 'collective politics' that avoids the violence of 'we' (the hierarchical 'we' of a Western definition of first-world feminism, whether second or third wave, essentialist *or* anti-essentialist), it is one which places a particular emphasis on labour: the 'painstaking labour' of getting closer to each other, working for each other and speaking to (not for) others in order to find out what, as women, 'we might yet have in common' (*Strange Encounters*, p.180). Since the task is to find out 'what we might yet have in common', Ahmed insists that any encounters in this process must be premised on a recognition of the *absence* of knowledge that would allow one to control the encounter or predict its outcomes, allowing for both surprise and conflict.

With all of this in mind, we have curated this collection with a view to enabling different feminist ecologies to 'speak' to each other: to seek out different generations of scholarship and feminism; to have those who practise, those who write and those who do both of these things, 'speaking' alongside each other. Within this variety common interests do start to emerge, but the ways in which they are understood, articulated and explored remains highly diverse, contradictory, perhaps even potentially antagonistic.

Characteristic of many contributions is a self-reflexive looking back in order to 'see'/ 'see into' the future. For example, Dee Heddon reflects specifically on the legacy of feminist consciousness raising, as symbolized by the slogan 'the personal is political', and as reflected in past and present generations of autobiographical performance-making. Sue-Ellen Case's theoretical engagement with 'feminist memories and hopes' takes us backwards and forwards across critical moments of thinking and theorizing feminism and performance in the interests of futures less oppressive than the past. In the context of black histories SuAndi argues the urgent case for histories that cannot afford to be left behind. Telling (performing) stories is not just a matter of creativity she argues, but of 'filling the voids of history', Chapter 8, p.125). Aoife Monks focuses on women directors in different theatrical environments, some of who might not define themselves as 'feminist' but whose work, she argues, nevertheless plays with gender roles in re-presenting the theatrical canon of the past in ways that allow possibilities for future (feminist) transformations.

A concern over the current problematics and the possibilities of the feminist 'we', runs throughout most contributions but is at the forefront of Janelle Reinelt's reflections on British and American feminist playwrights, framed so as to interrogate the appropriateness of the term 'feminist' in a historical, (post)feminist moment: in a contemporary context in which it is no longer clear what the term 'feminism' designates. In a collaborative contribution, practitioners Leslie Hill and Helen Paris, take the rejection of the term 'feminist' by younger women as the starting point for an examination of the relationship between their own identification with feminism and the work they produce in their theatre company, Curious. This work sometimes may be 'overtly feminist' but at others is 'more layered' with other contexts and concerns 'in which the feminist strands don't stand out as clearly' (Chapter 4, p.69). Elaine Aston explores work for the English stage by an up-and-coming generation of women playwrights to see whether the kinds of social and cultural issues dramatized can be instructive or insightful for a revitalization of a feminism that remains 'attached' to socially progressive, democratic principles.

Virtually all contributions, even when concerned primarily with the 'local', are also firmly and consciously situated against the background of current global politics. Very specifically, Joanne Tompkins's analysis of the use of the desert location in recent Australian drama by women dramatists raises key issues concerning the Australian 'political unconscious' as a means of underlining the importance of questions of subjectivity space and spatiality in the context of postcolonial globalization. Similarly, connecting local with global in the context of Asian British theatre and the arts generally, Meenakshi Ponnuswami focuses on the 'new' or 'second-generation' citizen to argue the importance of the arts to visualizing the international, to imagining transnational futures. In the context of today's 'new technological Eden' and 'the globally expanding industry of ART [Assisted Reproduction Technology]', Anna Furse draws on her own work *Yerma's Eggs* to explore the possibilities of new media and performance, and offers a timely feminist reminder that where 'the real power lies is not with the tools of such [reproductive] technologies themselves, but with those who control them' (Chapter 10, p.166). The collaborative Chapter 11 written between Elaine Aston, Gerry Harris and Lena Šimić reflects on their experiences in attending an international women's theatre festival in Denmark as a means of examining some key issues for feminism and theatre in a global context.

Finally, in many ways the cross-generational interview with four, very different, New York women artists – Lenora Champagne, Clarinda Mac Low, Ruth Margraff and Fiona Templeton – at the close of this collection, encapsulates on a micro-level the macro-aspirations of the whole. There is every need, as Lenora argues at the end of the interview, to 'know you're part of a community. That you aren't alone. That we are going to be able to withstand this time and go forward.' Whatever the limitations of this project, this book is our way of opening up the question as to whether or not an affective feminist 'we' (community) in theory, and in theatre and performance practice, still has possibilities as a place for the painstaking labour of 'getting closer', of trying to find out what as women 'we might yet have in common' – in the recognition that 'we' are even now at the very beginning of this labour, not at the end.

Notes

1 Sue-Ellen Case, *Feminism and Theatre* (Basingstoke: Macmillan, 1988), p.4.
2 Susan Faludi, *Backlash: The Undeclared War Against Women* (London: Vintage, 1992), pp.70 and 101–5.
3 See, for example, Janet Lee, 'Care to join me in an upwardly mobile tango' and Shelagh Young, 'Feminism and the politics of power', both in Lorraine Gamman and Margaret Marshment, eds, *The Female Gaze: Women as Viewers of Popular Culture* (London: The Women's Press, [1988] 1994) pp.166–973 and 173–89.
4 Natasha Walter, *The New Feminism* (London: Virago, 1999), p.4.
5 See Sara Ahmed, *The Cultural Politics of Emotion* (Edinburgh: Edinburgh University Press, 2004), ch. 8, 'Feminist Attachments', pp.168–90.
6 Hélène Cixous in *The Newly Born Woman*, Hélène Cixous and Catherine Clément, trans. Betsy Wing (Manchester: Manchester University Press, 1987), p.75.
7 Elin Diamond, "The Violence of 'We': Politicizing Identification", in Janelle G. Reinelt and Joseph R. Roach, eds, *Critical Theory and Performance* (Ann Arbor: The University of Michigan Press, 1992), pp.390–8. Diamond observes that in Judy Chicago's *Dinner Party* exhibition, out of all the plates representing great women only the Sojourner Truth plate lacked a vagina, and Ellen Moers's monograph *Literary Women* generally excluded black women (with the exception of Lorraine Hansberry), p.393.
8 Hazel V. Carby, 'White Woman Listen! Black feminism and the boundaries of sisterhood', in Centre for Contemporary Cultural Studies, eds, *The Empire Strikes Back: Race and Racism in 70s Britain* (London: Hutchinson, 1982), pp.213–35.
9 See bell hooks, *Feminist Theory: From Margin to Center* (Boston, MA: South End Press, 1984), p.18.
10 Gloria Anzaldua, *Borderlands/ La Frontera: The New Mestiza* (San Francisco, Aunt Lute Books [1987], 2nd edn 1999), p.231.
11 Stanlie M. James, ed., 'Introduction', *Theorising Black Feminisms: The Visionary Pragmatism of Black Women* (London and New York: Routledge,1993), pp.1–12, p.7.
12 Donna Haraway, 'A Manifesto for Cyborgs', in Linda J Nicholson, ed., *Feminism/Postmodernism* (New York and London: Routledge, 1990), pp.190–233.
13 Judith Butler, *Bodies that Matter: On the Discursive Limits of 'Sex'* (London: Routledge, 1993), p.2.
14 Like Haraway's thinking, Butler's ideas have had a profound impact on feminist theatre and performance practice and criticism largely because her ideas appear to pertain directly to these fields of cultural production (not least because when articulating strategies of 'subversive repetition' of the sex/gender system, she points to examples of (theatrical) drag performance).
15 Andy Medhurst, 'Camp', in Andy Medhurst and Sally R. Munt, eds, *Lesbian and Gay Studies: A Critical Introduction* (London: Cassells, 1997), pp.274–93, p.283.
16 Jane Flax, *Thinking Fragments: Psychoanalysis, Feminism and Postmodernism in the Contemporary West* (Berkeley: University of California Press, 1990), pp.56–7.

17 Isobel Armstrong, *The Radical Aesthetic* (Oxford: Blackwell, 2000), p.198.

18 Karen Finley, *The Constant State of Desire*, in Lenora Champagne, ed., *Out From Under: Texts by Women Performance Artists* (New York: Theatre Communications Group, 1990), p.61.

19 Susan Bordo, *Unbearable Weight: Feminism, Western Culture and the Body* (Berkeley and Los Angeles: University of California Press, 1995), p.224.

20 See Sara Ahmed, *Strange Encounters: Embodied Others in Post-Coloniality*, (London: Routledge, 2000), p.10.

21 Alan Sinfield, 'Diaspora and Hybridity: Queer Identities and the Ethnicity Model', in Nicholas Mirzoeff, ed., *Diaspora And Visual Culture*: *Representing Africans and Jews* (London and New York: Routledge, 2000), pp.95–114, pp.105–6.

2

Navigating Postfeminism: Writing Out of the Box

Janelle Reinelt

My focus is ostensibly British and American feminist playwrights, but what I intend to raise is the appropriateness of the term 'feminist' in a historical moment when the circumlocutions necessary to utilize this term render 'appropriate' usage difficult. Not only is there the question of whether or not playwrights self-identify as feminist, but also the responsibility of theatre scholars to characterize the contemporary context carefully and honestly. However, the 'contemporary context' is precisely one in which it is not clear what the term 'feminism' designates.

Reluctantly, I have reached the conclusion that we live in a time of postfeminism. There is something performatively defeatist about using the designation 'postfeminism' – defeatist in that it seems to give up on the project of feminism, and performative in that it actively constructs the present based on a sense of feminism as past or over. I fear that to speak about it thus might make it true, and dawdle at the keyboard, avoiding my own prose, as if the word had some magic effect belonging to childhood. The adult scholar in me acknowledges, as any discourse theorist knows, the power of rhetoric and repetition to eventually dominate hegemonic conceptual thinking, and thus the fear of the magic is not completely without basis. Yet I cannot dodge the uncontrovertible facts that 'times have changed'.

An overarching umbrella movement of organized social or political commitment does not exist now at the grass roots or national levels in either North America or Britain, unlike the heyday of second-wave feminism (roughly between 1970 and the mid-1980s). Local, individual groups and political campaigns exist, but the identification of women with a Movement – the 'Women's Movement' has virtually vanished. Some lesbian and queer women identify with the gay/lesbian/bisexual/transgender (GLBT)

Movement, which still has some identifiable status as a Movement, especially in North America, but the relation between feminism and this movement is itself historically troubled and in no case is the assumption of a cross-over justified today. NOW, the National Organization for Women, is just another North American political pressure group, and while not irrelevant to national politics in the United States, it is extremely weak and disunified. A 2003 study in *American Sociological Review* reports: 'There is no such dominant conception [of feminism] any longer. Rather, the outcome of the various contestations regarding the meaning of feminism may be a *decreasing consensus* among younger cohorts about what identifying as a feminist implies.'[1]

The lack of meaning of the term 'woman', so well interrogated by the intellectuals of the Second Wave, has meant that there is no mobilizing term under which heterogeneous women can identify and organize. By the mid-1990s, feminism had a hard time forming a focal point because identifying with any common properties held by all women seemed impossible. Not only had Julia Kristeva taught us that the signifier 'Woman' is a cipher under which real women cannot find a place, but also concrete criticisms of women of colour and lesbians that white, liberal, heterosexual women were inadvertently speaking for them combined with charges of cultural ethnocentrism to divide women sharply. As the notion of identifying with a group becomes less tenable, the mobilization of women as a voting block has become even more questionable. In the US election of 2004, women could be seen to have actually put Bush in the White House for a second term. The margin of difference between the number of women supporting Bush compared to those supporting Carey adds up to one reason why Bush won the White House.

Looking at generational differences, some of the criticisms of second-wave feminism persuaded young women that feminism was old fashioned or too rigid; for example, the charge of extremism, that feminism had become old fashioned and intolerant.[2] In Rene Denfeld's words, feminism was upholding

> a moral and spiritual crusade that would take us back to a time worse than our mother's day – back to the nineteenth-century values of sexual morality, spiritual purity, and political helplessness... Current feminism would create the very same morally pure yet helplessly martyred role that women suffered from a century ago.[3]

She refers here to campaigns against sexual harassment, date rape and pornography, associated with some forms of feminism. Along the same lines as Denfeld, Katie Roiphe and Naomi Wolf also claimed that feminism has portrayed women as victims and that this is self-defeating and untrue. These arguments against a perceived puritanical morality and a culture of victimization are persuasive in contemporary North American culture because of their intersections with certain entrenched traditions of libertarianism and self-sufficiency. Although having a long history of American Puritanism in its past, American ideology also contains an impassioned commitment to personal freedom and First Amendment rights shared by significant constituencies on the Left and also on the Right of the political spectrum. This phenomenon has produced odd allegiances between right-wing Christian groups and feminists on some issues like pornography, and has alienated the more libertarian feminists or gay/lesbian/bisexual feminists for whom the struggle for sexual freedom has been the highest priority.

The discourse of victimization must be seen against a larger tapestry of social injustice for poor, non-white people across society. How can middle-class women claim the status of victim when compared to the homeless or otherwise downtrodden? However, on the other hand, conservative politicians have successfully overthrown the appeal to injustice by treating victims as those who should be held responsible for their own situation, and have recoded 'victim' as a term only appropriately belonging to victims of crime. In death penalty arguments in the United States, one argument of the pro-death penalty people is that the victims have a right to witness the retribution for their injuries in the form of execution. Thus the rhetoric of victims and victimization is highly volatile. No one wants to be a victim, but if one is, retribution is the desired outcome. This complicates and strengthens the perception that feminists indulge in an unwarranted discourse of victimization.

Young women have moved toward a greater individualism and away from identity politics, disliking labels and seeing no need to organize. A survey of college women aged 18 to 34 confirmed this aversion:

> The vast majority of campus women surveyed (68 percent) noted they don't like being identified by labels that describe their political party, sexual preference, ethnicity, religion, race, or physical ability. They dislike labels because they believe they are inaccurate, stereotyping, and insensitive; the labels that define sexuality (straight, lesbian, bisexual, etc.) and 'feminist' grated the most.
>
> (Rowe-Finkbeiner, *The F Word*, pp.5–6)

These features – the lack of an energetic and robust political movement, the lack of an identity 'woman' with which to align, and the perception of restrictive and detrimental positions associated with feminism – have contributed to bring about the state of postfeminism, a state in which there is nothing to join and no clear 'woman' to be, but in which many of the concerns of actual women about equality, free expression, power, respect, and sexual subjectivity are still present and compelling. In the mid-1990s, for example, men were ten times as likely to hold office as a member of parliament – an average across all countries of the world. In first-world capitalist countries, men's average incomes are approximately double women's. R. W. Connell explains that 'the more familiar comparisons, of wage rates for full-time employment, greatly understate gender differences in actual incomes' because men are more likely to have shares of stocks and other financial dividends as a result of positions as head of corporations or major shareholders.[4] In 2002, women accounted for only 5.2 per cent of top earners at Fortune 500 companies. Thus the material conditions under which women live show that, using basic economic and political comparisons, women are still oppressed under patriarchy.

These concerns make up what I call the feminist residue from the Second Wave – serious issues have been identified and are still present, but they are ignored, pushed aside or simply denied. Talk of the progress of women tends to obscure how difficult the situation really is. Many of my young women students often show interest in women's studies or gender issues while simultaneously insisting with strong conviction they are NOT feminists. Their view seems to be that the need for feminism is over, and they are uncomfortable with the term, rejecting I'm not sure what, but I suspect the stereotyped media image of feminists as man-hating and bra-burning, the 1970s equivalent of the spinster of earlier years, the spectre of the unmarried and shrivelled up maiden aunt that second-wave feminism was at pains to deconstruct as a misogynist fiction of patriarchy. Ironically Women's Studies and Gender Studies have become a legitimate part of most university curricula in North America and Britain while the movement that made this possible has all but disappeared. Academics have been accused of forgetting their roots in activism, and the mounting pressure to increase research productivity in both American tenure and promotion practices and British research assessments has made it increasingly difficult to define community service, outreach or other quasi-organizational activities as legitimate tasks for the professoriate. Furthermore, the ideal situation, taken for granted during the Second Wave, combines activism and scholarship to fuel a genuine movement for social change – which always requires action and reflection in tandem to succeed.

The theatre world is similarly experiencing a decline in feminist dramas, although incorporating the significant success of individual women artists. In 2001–2, of 2000 off-Broadway and regional theatre productions in the United States, 16 per cent were written by women and 17 per cent had female directors. I suspect these figures would be the same or lower in the United Kingdom. So much for equality in the workplace.

Most of the feminist theatre companies, like feminist bookstore collectives, had disappeared in the United States by the mid-1990s. Not only is it always difficult for small theatre companies to survive on what amounts to no fiscal support from the government, it has become even more difficult to survive in recent years in a conservative climate and with less and less dollars available for arts funding. Similarly in Britain, the Thatcher years took a heavy toll on left-wing and feminist theatre companies, perhaps the most well known casualty of that era being Monstrous Regiment, who for over 20 years had managed to thrive and produce many pieces created by and for women with explicit socialist feminist politics as their artistic mission. Solo performers have perhaps managed to survive the best – one thinks of Bobby Baker in Britain or Karen Finley or Holly Hughes in the United States, and most prominently, of Anna Deavere Smith, whose solo work has been widely seen in the mainstream through television broadcasts of her work and Broadway and major regional productions at home and abroad.

In Britain, some women playwrights have remained well established through the 1990s. Caryl Churchill and Timberlake Wertenbaker, for example, are feminist writers from the Second Wave who still write and are produced in major venues such as the National Theatre and the Royal Court. As for directors, Ann Bogart or JoAnn Akalatis in the Unites States and Deborah Warner and Di Trevis in the United Kingdom have achieved a great deal of critical acclaim and work in the major professional theatres in both countries. 'Feminist', however, would not be a descriptive term brought easily to their work, although two of them might self-describe as feminists in their 'private' life. Only Warner consistently thematizes gender issues in her directorial work, and her perspective is rendered complex by her well-known lesbian partnership with Fiona Shaw, with whom she has created many projects, such as the cross-dressed *Richard II*. These women are now all significantly senior. In Nicholas Hytner's semi-miraculous burst of vitality at the National during his new regime, he has chosen few women as part of his inner cadre, and Katie Mitchell, arguably one of the insiders, is no longer a 'youngster', now herself aged over 40.

One of the early concerns of feminist representation was to get women in central roles represented as legitimate protagonists of dramas – not just in order to create more roles for women, but to treat their lives, in a gloss on the words of Erich Auerbach in *Mimesis*, as serious subjects, to ensure 'the rise of more extensive and socially inferior human groups to the position of subject matter for problematic-existential representation'.[5] Along with this objective came the reclaiming of women's history. To some extent, these features continue to be present in the contemporary work of most of the women artists mentioned above, but they may not be as central or as sufficient as they once were. Di Trevis's stunning production of Proust's *Remembrance of Time Past* (2001), for example, foregrounded a gay male imagination, but placed no particular stress on female representation or feminist features. Ann Bogart's *tour de force, Bob,* is a solo piece about Robert Wilson. It goes a long way to explaining a great deal about this theatrical artist and his work, but obviously has nothing to do with women, nor are there any on the stage.

In both countries, North America and Britain, some individual artists and groups continue to survive on the margins, playing to faithful and enthusiastic audiences in small venues or touring sites. Often these are identified as lesbian or queer performers rather than feminists. Split Britches, an American lesbian company, and Bloolips, a British gay company, for example, have collaborated productively across the gay/lesbian and the US/UK divide, appearing at the WOW café on New York's lower east side or at The Drill Hall in London's fringe. Bobby Baker has a large following in the United Kingdom, and in the United States, Lisa Kron has a viable solo performance career that started with her work as one of the Five Lesbian Brothers. Thus the residue of second-wave feminism exists in various facets of mainstream and subcultural performances. To ignore it is to deny both its history and its present state of play. On the other hand, the work mostly remains on the margins, and is not any longer as strongly identified with feminism as it would have been a decade ago (Bobby Baker is an exception as she continues to work within an explicitly feminist tradition).

Turning to playwrights, we find some British and American women for whom the claims of the residue of feminism, if not more, can be made. By examining their work through the lens of longstanding feminist concerns, we can come closer to understanding the important legacy of second-wave feminism, and the way in which today's women grapple with sex and gender even when these are not thematized directly in their work.

Caryl Churchill is the most straightforward well-known feminist writer. Her recent work from *The Skriker* to *Far Away* at least still involves women at the centre of the action, and is still concerned with the quality of women's lives in contemporary society. Churchill is the great icon of second-wave feminism in the British theatre because she was a conscious and insightful commentator on it during the 1980s, and because her best known plays, *Cloud 9, Top Girls, Vinegar Tom, Fen* and *Serious Money,* became defining texts for theatre scholars seeking to describe what constituted feminist performances in the theatre. Using a British epic form familiar from Bertolt Brecht's work but focused on gender and sexuality as well as economic and social issues, Churchill became expert at the feminist gestus, what Elin Diamond has described as those moments in the play text 'when social attitudes about gender and sexuality conceal or disrupt patriarchal ideology'.[6] In *Cloud 9,* for example, Churchill embodied the Victorian styles of gender on cross-dressed bodies to reveal gender as an historical construction under pressure, and in *Vinegar Tom,* she explored the complicity of the state and religion in the production of women as perceived deviants and witches because they threatened the stability of the economy of seventeenth-century England. In *Top Girls,* she took on a feminism corrupted by Thatcherism, indicting individual economic success without socialist commitment, and continued this analysis in *Serious Money,* her comic yet savage portrait of financial traders and stock jobbers in Thatcher's second term. These plays managed to succeed in major venues such as the Royal Court Theatre, Britain's premiere theatre for new writing.

Churchill still enjoys the prestige of identification as one of Britain's most important dramatists, but her work is changing. *A Number* (2003) is about cloning, or rather, the demise of identity based on patriarchy. An extremely interesting and timely play, it has no women in its cast, and its topic is a crisis in masculinity brought about by interrupting and changing kinship lineages in the wake of the possibilities of cloning.

Even *Far Away,* which premiered at the Royal Court Theatre in November 2000, is far from being exclusively about women's issues. Set in a world of some undisclosed future, it appears to be about species alienation and warfare, where all God's creatures are divided into warring camps. One of the characters explains who the bad guys are: 'Mallards are not a good waterbird. They commit rape, and they're on the side of the elephants, and the Koreans. But crocodiles are always in the wrong.'[7] In this Brave New World, conflict is ever-present and loyalties keep changing. The one overtly feminist through-line is the development of the character of the little girl of the first scene, Joan, who raises questions about immigrants who are being beaten by her uncle. In the course of the play, she grows up into a world of global species warfare. Consistently through all of Churchill's work, and often figured specifically as a child character, there is a concern for the experiences and future of young girls. What kind of world will they inhabit and inherit? might be her constant question.

In the play, Joan takes a job making hats. She and her co-worker Todd, later her husband, take pride and pleasure in making them for a competition culminating in a parade. Joan has trained to make hats in college. Joan and Todd talk about work conditions in the firm, and Todd is determined to speak up for improvements, even if it means risking his job; Joan admires him for this. Eventually, we find out they are making hats for prisoners, for 'ragged, beaten, chained prisoners', to wear on their way to execution (*Far Away*, p.24). Joan and Todd do not even discuss this, however. It is as if they are ideologically blinded to the uses for which the hats are being made. Churchill is surely lodging a critique against the transnational mode of production in which aesthetics are used to occlude material suffering elsewhere. As this scene is also a heterosexual courting scene, it is also about how a young woman is socialized into work and sexual attachment through the new transnational frame.

Is this enough to count as a feminist text? Or is the concern with globalization, clearly the heart of her play, separate from feminism? I am myself divided on this question. I think, given Churchill's full body of work, it is relatively easy to place this play within the corpus and see it as inflected with feminist politics. On the other hand, I think Churchill is addressing some other issues in this work, issues that have come to prominence recently as a result of global expansion and the consolidation of world markets, and that gender is taking an appropriate back seat here to the question of ideological blindness based on the ability to dissociate from the suffering of others, threatening to lead to an apocalyptic species warfare.

The second British writer of major prominence I want to consider is Sarah Kane. Her last play, *4.48 Psychosis*, has been widely produced, including a US tour in 2004 of the definitive production directed by James McDonald. Kane is generally recognized as a great talent; the power of her writing compares to Edward Bond and even Samuel Beckett. She was born in 1971, just at the start of second-wave feminism, and she died of suicide in 1999. Young, gifted and dead. Certainly not a feminist, but a woman writer. What are we to say of her? Her plays have been received as shocking because of her depictions of brutality and violence and the seeming lack of redemption in her imaginary universe. Critics puzzle about the nihilism of her plays (whether they really are nihilist or not), and about the style of her expression – poetic, anti-realistic, dramaturgically unique. They have not written about her much in terms of feminism, although as this essay goes to press, a number of graduate students are working on dissertations on Kane.[8]

Blasted is often described as Kane's response to Bosnia – not literally, not citationally, but depicting the situation of violence and the relative disengagement of the West. She connects rape, the rape of a young women in a hotel room, to a scene of civil war. The innocence of Cate contrasts with the violence and madness of her lover Ian. Later the contrast extends to a soldier who appears suddenly 'blasted' into her world. This is not a play about gender essentialism, but its categories fit that epitaph. Cate doesn't eat meat, doesn't believe in killing, isn't racist – unlike her lover. Ian is suffering from some physical internal malady, perhaps cancer, which keeps him in a state of protracted pain while his job as a hired killer keeps him in a state of anxious paranoia. When the hotel scene is literally destroyed by a blast and shifted to a war zone, male violence and capacities for evil take over the narrative as rape and torture are described and enacted. This play, as is common in Kane's work, asks if there are any limits to human cruelty and seems to answer 'no'. However, Cate tries to save the baby she retrieves from the war, and in the closing moments, shows compassion for Ian in spite of his previous brutal treatment of her.

The storm of controversy over *Blasted* when it was first produced in 1995 is now a well-known story.[9] The staging of savage acts – the soldier sucks out and eats Ian's eyes and Ian eats the remains of the dead baby – seemed sensationalist and brutal to reviewers. Her playwright colleague, David Greig, writes in the introduction to her collected plays:

> [h]er simple premise, that there was a connection between a rape in a Leeds hotel room and the hellish devastation of civil war, had been critically misunderstood as a childish attempt to shock. It would not be until *Crave* in 1998 that public perceptions of her work would begin to move beyond the simplistic responses of the *Blasted* controversy.[10]

Looking back on *Blasted* now, it seems remarkable that no sustained discussion of the representation of gender in the play emerged. The play locates behaviour not exclusively in biology but in cultural experiences, making clear the gender entailments of masculine anxiety and rage. As feminist psychologist Nancy Chodorow writes in her study of the psychodynamics of masculine violence:

> Men, after all, are directly responsible for and engage in the vast majority of both individual violence and rape as well as collective violence. Historically and cross-culturally, they make war. Men are soldiers and, as politicians and generals, those who instigate and lead the fighting. Men also engage in extreme violence: they are (mainly) the concentration camp guards, the SS, those who perpetrate genocide, mass ethnic rape, pogroms, torture, and the murder of children and old people. Hormones, the structure of masculine personality, and/or the social and political organization of gender, male bonding, and male dominance, all lead many men to react to threats with violence and aggression in a way that most women do not.[11]

Why not see Sarah Kane as representing this unfortunate truth about historical gender in her play? She does not need to be a self-proclaimed feminist or a biological essentialist to have created a dramatic parable that captures this state of affairs.

In a career that lasted just five short years, Kane continued to address the limits of cruelty, the limits of misery, and the fragmentation of selves caught in the crucibles of suffering. Gradually the inner hell of depression and madness became her dominant topic, but always with the recognition that these inner states were product of, and developed through, the social relations surrounding them. Her concern was never just with women; most tellingly, in her last two plays the characters are not identified as gendered, and various combinations of casting become possible, producing different gendered effects. Her last play, *4.48 Psychosis* is directly about depression and suicide, and it is almost impossible to resist seeing it as her last words, since she finished it shortly before her death and dealt explicitly with the topics of depression, psychotherapy and chemical treatments that marked her own unsuccessful struggles with depression. Consider this excerpt:

> I can't think/I cannot overcome my loneliness, my fear, my disgust/I am fat/I cannot write/I cannot love/My brother is dying, my lover is dying, I am killing them both/I am charging toward my death/I am terrified of medication/I cannot make love/I cannot fuck/I cannot be alone/I cannot be with others/My hips are too big/I dislike my genitals/At 4.48 when desperation visits I shall hang myself to the

sound of my lover's breathing/I do not want to die/I have become so depressed by the fact of my mortality that I have decided to commit suicide/I do not want to live.

(Sarah Kane, *Complete Plays*, p.207)

This passage is full of the very sort of self-loathing expressions second-wave feminism tried to fight – preoccupation with the imperfect body, lack of self-image or sense of self-value, depression, guilt, masochism.[12] The typical narrative for women in drama, feminists had observed, is transgression followed by death – thus Hedda Gabler, Marie in *Woyzeck*, all the opera heroines, foremost perhaps Carmen. 'We must stop representing death as inevitable for women who dare to be different', we feminist critics thought. And here is Sarah Kane, who makes a definitive performance of her life, by creating her art and doubling it in her tragic actions. She willed it, she chose it (yet she didn't), and she seemingly provides an example of the lack of escape from the deep desolation of the feminine psyche. Should I call her feminist? Do I have the right? The *via negativa* of her example is a protest. Kane was angry at the world as well as at herself. She demanded her audiences pay attention to real suffering, her own but also others. Can we not see the residue of feminism, its ethical and political commitments, even in her seeming rejection of it? Is the struggle with fragmentary identities, the desire for love and intimacy, the experience of deep alienation not both a feminine struggle AND a differently sexed and gendered struggle in her work? She has represented same-sex love, heterosexual love and need, male and female grief. Her tapestry is large enough to apply not only to women, but her own personhood insists on the gendered nature of her experience and, finally, I would argue, her art.

Yet there is an additional ethical issue here. Kane was from a different generation after all. She did not want to be associated with feminism. What happens when someone like me stakes a feminist claim for her work – even a residual claim. Is it a kind of betrayal... perhaps even of sisterhood?

The third playwright I wish to consider is Suzann-Lori Parks, an African-American playwright who received the 2004 Pulitzer Prize for her play, *TopDog/UnderDog*. Some of her plays have seemed very close to second-wave feminist work – I am thinking specifically of *Venus*, her 1996 play about the historical figure Saartjie Baartman, the 'Venus Hottentot', who was exhibited in side-shows in London in the nineteenth century because of her large buttocks. In her play, Parks raises questions of representation (both how Baartman was represented historically and how to represent her today). Through a series of epic, Brechtian scenes, fragments of supposed historical documents, such as her baptismal record, the autopsy report after her death, or the trial record considering if Baartman should be allowed to be displayed in the circus, the drama takes on a quasi-documentary style, complicated by lyrical, surreal language and images that give the drama a nightmare quality. Care is taken with the representation of Baartman not to fetishize the role by reinscribing the status of fetish object in the actress. Saartjie sometimes appears in a padded suit to avoid realistic portrayals that would invite voyeurism by providing an unmediated object for a male racialized gaze. Parks offers what amounts to a socialist feminist analysis of the economic and racial causes of Baartman's exploitation, combining race and gender in a devastating critique of white capitalist and masculinist culture, while inserting Baartman firmly into history.

Yet this play is not typical of her work, which has ranged over many topics and expressions, including *The America Play* and *The Death of the Last Black Man in the Whole Entire World*, both of which feature only supporting female characters and treat men as explicitly central. *The America Play*, her most profound and accomplished text, is the precursor to *Top Dog/Under Dog*. Its main character impersonates Abraham Lincoln in a sideshow at which people pay to play Edwin Booth, Lincoln's assassin, and re-enact his shooting. This character is called 'The Foundling Father', with its play on foundational status and orphaned status. The scene, Parks tells us, is 'A great hole. In the middle of nowhere. The hole is an exact replica of The Great Hole of History.'[13] Self-consciously playing fast and loose with the past, Parks asks what it would take to create a meaningful history for African-Americans by excavating the archaeology of national myth.

In *TopDog/UnderDog*, the main characters are brothers, the sons of the Foundling Father. They are engaged in a fraternal, competitive struggle around identity and kinship. As in Churchill's *A Number*, the play has no female characters at all, although African American performance scholar Harry Elam has remarked that the play ironically contains a female voice in its absence and silence, which is certainly true of Chruchill's *A Number* as well.[14] The gaze, we might say, can be feminine in these plays.

While I might argue that the majority of Parks's work is not accurately characterized as marked by feminist concerns, she herself reminds us of another aspect from which to view her author-function. Parks commented in an interview, 'Just being an African-American woman on Broadway is experimental. As far as I know, there are only four of us:

Lorraine Hansberry, Ntozke Shange, Anna Deavere Smith, and me. It's also experimental to write a play that just involves two men and to write it so well that people think a man wrote it!.'[15] Thus while Parks does not identify as a feminist, she is certainly quite aware of writing as a woman in the American marketplace. Indeed, many African-American women do not identify with feminism, believing that the Movement was mainly about white middle-class women, while still being extremely aware of their material and symbolic conditions in a society that often makes them doubly oppressed on account of sex and race. These women write about many things, including gender, but they do not always fit comfortably into generalizations about feminist writers. Along with Parks, I am thinking of other African-American women writers such as Pearl Cleage, Kia Cothrun or Adrienne Kennedy. As with Sarah Kane, there is an ethical issue involved if feminist theatre scholars like me try to insist too strongly on an affiliation that the writer herself disparages.

The final playwright I want to discuss is American playwright Paula Vogel. In this case, Vogel does identify as both a feminist and a lesbian, but I want to discuss a play of hers that has upset some feminists and, in addition, seems to have nothing to do with lesbian representation. *How I Learned To Drive* premiered in 1997 and received a Pulitzer Prize. It has been widely performed throughout the United States in regional theatres. Its success in the larger theatrical mainstream raises questions about its politics, because in order to survive and do well in the commercial American theatre (Broadway and the large regional circuit), plays by and about women are frequently extremely hegemonic and covertly conservative. In the 1980s, Beth Henley, Marsha Norman and Wendy Wasserstein received Pulitzer Prizes for plays that, as feminist newspaper critic Alisa Solomon observed, represent 'intelligent, educated women, and assure us that they are funny for the same traditional reasons women have always been funny: they hate their bodies, can't find a man, and don't believe in themselves'.[16] Vogel's play, then, is somewhat suspect in its success. It treats incest and childhood sexual abuse, told to us through the narrative of Li'l Bit, a 40-year old who takes us back and forth chronologically through the phases of her adolescence when she was involved in an incestuous relationship with her Uncle Peck. He taught her to drive, thus the title, as well as initiating her into sexual desire. The play, however, is not simply a trauma narrative which releases and elevates its victim through the process of the telling. It is rather a complicated document that represents the ambivalence of Li'l Bit toward Uncle Peck, and while not excusing his sexual abuse, acknowledges in the course of the action both his complex character and Li'L Bit's reasons for investment in their relationship. Reminiscent in subject matter of *Lolita,* but not told from the male point of view, *How I Learned to Drive* is troublesome to certain viewpoints in contemporary feminism. As performance scholar Ann Pellegrini writes:

> Feminists – both clinicians and non-specialists – fought to expand the category of trauma and 'post-traumatic stress disorder' (PTSD) from its focus on 'masculine' injuries like shellshock or war trauma to include such 'feminine' injuries as incest and rape. This victory has not come without cost, however... specifically where incest is concerned, the PTSD label leaves out of the view the ambivalence that characterizes the relations of many 'victims' to their 'perpetrators.' There is a complexity and variability to victim/perpetrator relations that PTSD cannot lay hold of or recognize.[17]

In the opening moments of the play, Li'l Bit tells us: 'Sometimes to tell a secret, you first have to teach a lesson.'[18] Pellegrini argues that the secret is not the incest, which is pretty clear from the opening scene of the play, but rather the ambivalence and identification that Li'l Bit feels for Uncle Peck. It has to remain a secret until the narrative of the complexity of their relationship unfolds. David Savran, a gay performance scholar, has commented that he sees Vogel's work as dramatizing women who 'sustain powerful psychic identification with men'. Seeing Li'l Bit as playing a masculine role relative to driving the car, Savran sees that Li'l Bit learns 'to be both a desiring subject and desired object, both molester and violated one, both initiator and disciple, both taker and giver' (Savran, *A Queer Sort of Materialism,* p.195). For these reasons, the play is very controversial – among feminists particularly, for whom Uncle Peck seems to be getting away with something, and for fear that spectators in the theatre might seem to be witnesses taking pleasure in incestuous representation. For in order to represent Li'l Bit as ambivalent about the sexual feelings she's experiencing, the audience must be allowed that ambivalence as well. Or so the argument goes. Pellegrini and Savran, in defending Vogel's project in the play, are attempting to address the very feminist and queer spectators who may be raising these objections. Here, then, is an interesting predicament: there is no homosexuality in the play, and so Vogel's lesbian identity is not overtly present in the representation. Since she is not making a straight PTSD analysis of Li'l Bit as a victim, she does not appear as a feminist writer.

Yet arguably, this play is explicitly addressed to feminists and queer folk in order to grapple with the received categories of current representational thinking. This is what I think makes *How I Learned to Drive* an important play for feminism.

What I am suggesting, of course, is that while these plays seem to lack the overt marks of explicitly feminist commitment, they are informed by, and filtered through, the perspectives of women who have been familiar with, and lived in relation to, second-wave feminism. Even Sarah Kane knew the themes and preoccupations of second-wave feminism. She sometimes satirized them, as in *Phaedra's Love*, and sometimes charted their inadequacy to capture real lives of suffering, as in *Blasted* or *Cleansed*. Suzann-Lori Parks does not consider herself feminist, and pursues what she must define as 'larger questions' of race and nation at the centre of her work, but when she writes a play like *Venus*, she reveals how well she knows the socialist feminist critique and adapts it for her own purposes. Churchill continues to identify as feminist, but her work is becoming more diffuse, less epic and more lyrical, and more removed from the familiar feminist writing of *Cloud 9* or *Top Girls*. Thus while none of this work is overtly feminist, it is imbued with second-wave residue, the presence of the 'post', the effects of coming after and of having encountered second-wave feminism. And, I would submit, this is not a negligible legacy.

In concluding, I turn to a new writer who shows continuity and change in a younger generation. Stephanie Dale turned 30 while she was working on 'What is Missing From Your Life?'. Based on interviews and letters with 200 women in Birmingham to whom she posed this question, Dale created a fictional frame within which to organize the wealth of diverse self-reflections this question elicited. A young newspaper reporter covers a story about an elderly lady who put up a placard in a church posing the question: What is Missing From Your Life?. Irene tells Helen she was amazed at the outpouring of letters she received and doesn't quite know what to do with them.

Dale's play was broadcast on BBC Radio on 14 September 2004. Radio drama, of course, was the medium second-wave feminists such as Caryl Churchill and Michelene Wandor and many others cut their teeth on three decades ago. Radio is a personal medium, focused on interiority, and carried largely by the voice. Dale's play perfectly made use of the strengths of this medium as the ages and classes and cultures of her subjects were intimately portrayed by four actresses. Part fiction, part documentary, the women's responses to the question range from the quotidian 'I always wanted to travel and I never have', to soulful laments for the serious and the trivial: 'I'm missing...I used to have dreams – dreams. And pretty hats!' Young women yearning for love contrast with older women dealing with the losses of their partners, mothers or friends; a woman who is probably Asian talks about the difficulty and necessity of respecting elders in her culture while a working-class woman complains about missing meaningful communication with her husband. The voices are intercut with each other and with a musical score in such a way that sometimes the women seem to be in conversation with each other, and at other times seem to be leading parallel lives. In the fictional frame, Helen is having a relationship problem with her photographer husband Alan. She has a chance to move to Glasgow for a new job; he doesn't want to move. She is clearly employed at a higher level than he is. His constant attempts to win her attention and reassurance backfire. He makes friends with Irene when photographing her, and when Irene falls and is taken to the hospital, Alan is the one who is called. Slight and inconclusive, the frame material represents how people's lives touch each other through chance or circumstance, how women are connected to each other and yet isolated, how all of us, including Alan, yearn for something more. In her interview with Helen, Irene confides, 'I want to be able to give them what they want, can you understand that?' Perhaps this impulse is the place where feminist and postfeminist perspectives touch.

Portions of this essay were first developed during my guest-professorship in gender studies at the University of Mainz in 2002, and subsequently appeared in a volume edited by Renate Gahn through Wissenschaftlicher Verlag, Berlin, in 2005. In addition, portions of it appear on the web at http://www.barnard.edu/sfonline/ps/reinelt.htm.

Notes

1 Quoted in Kristen Row-Finkbeiner, *The F Word: Feminism in Jeopardy* (Emeryville, CA: Seal Press, 2004), p.93.

2 See, for example, Camille Paglia, *Sexual Personae* (New Haven and London: Yale University Press, 1990), Katie Roiphe, *The Morning After* (London: Hamish Hamilton, 1993), Rene Denfeld, *The New Victorians* (New York: Warner Books, 1995), Naomi Wolf, *The Beauty Myth* (London: Vintage, 1991).

3 Quoted in *The Routledge Critical Dictionary of Feminism and Postfeminism*, ed. Sarah Gamble (New York and London: Routledge, 2000), pp.46–7.

4 R. W. Connell, *Masculinities* (Berkeley: University of California Press, 1995), p.82.

5 Erich Auerbach, *Mimesis: The Representation of Reality in Western Literature*, trans. Willard R. Trask (Princeton: Princeton University Press, 1953), p.491.

6 Elin Diamond, *Unmaking Mimesis* (London and New York: Routledge, 1997), p.54.
7 Caryl Churchill, *Far Away* (London: Nick Hern Books, 2001), p.33.
8 Ken Urban at Rutgers University and Alicia Tycer at University of California Irvine are two graduate students writing about Sarah Kane.
9 See Graham Saunders's account of its reception in *'Love Me or Kill Me': Sarah Kane and the Theatre of Extremes* (Manchester: Manchester University Press, 2002).
10 David Greig, 'Introduction', Sarah Kane, *Sarah Kane: Complete Plays* (London: Methuen, 2001), p.x.
11 Nancy J. Chodorow, 'The Enemy Outside: Thoughts on the Psychodynamnics of Extreme Violence With Special Attention to Men and Masculinity', *Masculinity Studies and Feminist Theory: New Directions*, Judith Kegan Gardiner, ed., (New York: Columbia University Press, 2002), pp.235–260, pp.251–2.
12 Note, however, this could be spoken by a male.
13 Suzan-Lori Parks, *The America Play and Other Works* (New York: Theatre Communications Group, 1995), p.158.
14 Personal conversation with Elam, 16 December 2003.
15 Iris Fanger, 'Pulitzer Prize winner shakes off labels', *Christian Science Monitor*, 12 April 2002: 19.
16 Quoted in David Savran, *A Queer Sort of Materialism: Recontextualizing American Theater* (Ann Arbor: University of Michigan, 2003), p.202.
17 Ann Pellegrini, 'Staging Sexual Injury: *How I Learned to Drive*', in Janelle Reinelt and Joseph Roach, eds, *Contemporary Theory and Performance* (Ann Arbor: University of Michigan Press, 2006).
18 Paula Vogel, *How I Learned to Drive*, in *The Mammary Plays* (New York: Theatre Communications Group, 1998), p.7.

3
Citizenship and Gender in Asian-British Performance

Meenakshi Ponnuswami

This begins with two premises: that a radical and utopian rethinking of the idea of globalization must be the promise and goal of feminism, its future; and that theatre and other performative arts have a key role to play in visualizing the international, making it possible to imagine bodies in transnational and transcultural interplay. Like Gabriele Griffin's important study of black and Asian women dramatists,[1] my discussion here is indebted to Avtar Brah's articulation of the transitional space of the global as a 'diaspora space':

> Diaspora space is the point at which boundaries of inclusion and exclusion, of belonging and otherness, of 'us' and 'them' are contested. My argument is that diaspora space as a conceptual category is 'inhabited', not only by those who have migrated and their descendants, but equally by those who are constructed and represented as indigenous. In other words, the concept of *diaspora space* (as opposed to that of diaspora) includes the entanglement, the intertwining of the genealogies of dispersion with those of 'staying put'. The diaspora space is the site where *the native is as much a diasporian as the diasporian is the native.*[2]

An important link between feminists of different cultures in the production of this utopian 'entanglement' is the 'new citizen': not the migrant, although her role in the dialogue is also important, but her children, 'second-generation' or 'new' citizens who are born or raised in the new homeland. Second-generation citizens are uniquely positioned to be agents of intercultural communication and exchange, as (at least in theory) they are acknowledged as reliable informants and translators by both natives and migrants. They are also uniquely capable of transforming adopted homelands into 'diaspora spaces'. Traditionally, claims of national belonging have been given legitimacy by claims to history – in Eric Hobsbawm's words: '[n]ations without a past are a contradiction in terms. What makes a nation *is* the past.'[3] Unable to claim the nation's history as her own, the new citizen devises alternative ways to perform citizenship, disinheriting nationalities and histories to produce what Stuart Hall calls a 'new ethnicity' that 'retheorize[s] the concept of "difference"'.[4] The new citizen's conflicting claims to nation-state and ancestry have the potential to create wholly new discourses of ethnicity and identity, and to perform citizenship and belonging in entirely new ways.

Ideally, quests for democratic rights by first- as well as second-generation immigrants would entail a complete rethinking not only of the national, ethnic and racial constitutions of the state in question but also of the notion of statehood itself. However, this sort of reconfiguration is usually forestalled by a strategic liberalization of the state's self-conceptualization. On the one hand, alien bodies are acknowledged as citizens while, as Nira Yuval-Davis puts it, 'the differences of class, ethnicity, gender, etc., [are considered] irrelevant to their status...'.[5] Then, often in response to a crisis such as race riots, 'difference' is acknowledged and accommodated by the state, for example in the form of subsidy for ethnic arts or grants to foster multicultural curricula in schools. Inevitably, this sort of accommodation into the cultural apparatus leaves the dominant ethnic core undisturbed, while placing ethnic minorities who seek citizenship in the odd position of demanding a legal status which in practical terms relegates them to the cultural periphery, singing and dancing from the margins in a kind of state-sponsored ethnic sideshow. A number of commentators have observed that, in Britain in the 1960s and 1970s, racism in the dominant culture encouraged communities of women to form political alliances across racial, ethnic and national boundaries, while by the mid-1980s and 1990s alliances across these divisions had become more complex and difficult to resolve. Pratibha Parmar despaired in 1987 that 'many women have retreated into ghettoised lifestyle "politics" and find themselves unable to move beyond personal and individual experience'.[6] Amina Mama similarly argued that 'a growing focus on identity and a new competitive cultural politics replaced the 1970s/early 1980s notions of black unity and wider anti-imperialist and black liberation struggles'.[7]

The difference is, arguably, not only an ideological but also a generational one. Earlier immigrant generations had confronted more unified forms of racism and were thus able to form coalitions based upon recognizable common interests. But later 'second-generation' citizens, facing questions about ethnic and national identity and affiliation that required a different set of analytical tools and new modes of self-definition, reinvigorated debates around nativity and belonging.

As Mary Karen Dahl puts it in the context of her discussion of black British theatre, 'self-articulation [became] a permanent necessity'.[8] In 'first-generation' British immigrant theatre, assertions of national belonging, of Britishness, are often pitted against nostalgic affirmations of an alien but inalienable heritage, producing an apparently irreconcilable conflict between the claiming of citizenship and the longing for history. But this is not usually the case with second-generation theatres, which have more typically repudiated essentialist ethnic or national affiliations and identities, and sometimes even history. The cultural contexts invoked by the theatre of second-generation black and Asian British women are global and historicized, addressing what Paul Gilroy calls the need to 're-conceptualise the whole problematic of origins'.[9] There are performances of retrieval – invocations of diasporic histories in narratives of the coming-of-age of new citizens – in which otherwise private rites of passage address wider political issues. This is certainly the case with plays as dissimilar as Winsome Pinnock's *A Hero's Welcome* (1989), Jackie Kay's *Chiaroscuro* (1986) and Trish Cooke's *Running Dream* (1988). In these and other history and memory plays, particularly, fractured and subjective memories are placed in the wider context of immigrant and racial history in such a way as to deconstruct notions of statehood and identity, as well as the processes of displacement, settlement and cultural mutation.[10] This is the performative rendering of Brah's 'diaspora space', memorably caught in the irreverent, hopeful image at the end of Hanif Kureishi's film *My Beautiful Laundrette* (1985), when the former Pakistani and the former National Front supporter make love in a renovated laundromat.

I am concerned, however, by some other manifestations of the new Britishness. It is possible to argue that the retreat into 'lifestyle' and identity politics led, perhaps unwittingly, from an anti-imperialist orientation to the development of a more *national* focus – what could in some cases be called a form of postcolonial parochialism. I am thinking of artists who came of age after Brixton and by the 1990s had begun to produce an original and exciting body of creative work, of the sort documented in Kobena Mercer's *Welcome to the Jungle* and Catherine Ugwu's *Let's Get It On*. Mercer argues that this 'new generation of black British cultural practitioners ... are in the process of creating new, hybridised identities', in part by undertaking a very foundational interrogation of identity formation, of 'the need for nation' itself.[11] But consider the terms on which Mercer describes this project:

> cinema and image-making have become a crucial arena of cultural contestation today – contestation over what it means to be British; contestation over what Britishness itself means as a national or cultural identity; and contestation over the values that underpin the Britishness of British cinema as a *national* film culture.
>
> (*Welcome to the Jungle*, pp.4–5)

At issue here is not specifically the forging of a new transnational – or transitional – subjectivity, but the remapping of the received contours of Britishness. Hybridized, postmodern and transnational identities do lurk behind second-generation quests for citizenship and equal rights, but, in asserting the Britishness of the black or Asian citizen, some younger performers and writers not only lose sight of the earlier struggles against imperialism and fascism (as Mama and Parmar had lamented), but also renew stereotypical and reductive images.

In this essay, I focus on performance texts by younger British Asian writers. It is important to consider film and television as well as theatre in this context, because these have provided some of the most important examples of ethnic imaging, and have been increasingly more influential than theatre in generating popular images of postwar and postcolonial British multiculturalism. Moreover, from the 1990s onwards, as opportunities for playwrights have dwindled generally, most writers and actors have worked simultaneously in film, television and theatre.[12] Indeed, many of the most prominent and critically acclaimed works by British Asians are novels, one of which I consider briefly below.

Like Kureishi's, some of these works do indeed in their playfulness have a carnivalesque, even transgressive edge in their depictions of South Asian ancestry: consider the satire of the Bombay film industry and the night-clubbing of the aunties in *Bhaji on the Beach* (1994), or Ayub Khan-Din's witty, raw, *East Is East* (1996). While such representations of generational conflict are far from respectful, they tease out the nuances, contradictions and comedy of first-generation hankerings for a lost authenticity as well as bungling second-generation performances of interculturalism. The liberational potential and limits of comedy are self-consciously captured in a poignant moment in *East Is East*, where the 'obedient' son Abdul recounts his first, defiant visit to a pub:

> We were sat drinking, telling jokes, playing music, telling more jokes. Jokes about sex, thick Irish men, wog jokes, chink jokes, Paki

> jokes. And the biggest joke was me, 'cause I was laughing the hardest. And they laughed at me because I was laughing. It seemed as if the whole pub was laughing at me, one giant grinning mouth. I just sat there and watched them, and I didn't belong...[13]

As Khan-Din is well aware, the willingness to reappropriate a stereotype may signal cultural poise, but it also entails a new kind of minstrelsy: as Samina Zahir comments of the television comedy *Goodness Gracious Me*, 'While for some [viewers] it shows the increasing self-confidence of British-Asians, for others it allows the white English yet another opportunity to laugh at the UK's Asian communities.'[14]

When parody fails to create a new image or insight, it simply reinstates the conventional binary of East and West. Transnational discourses of this sort problematize the received categories of British experience, but do so by deploying very worn notions about the ways in which Britishness and Asianness are performable. This is partly a function of their emphasis on 'identity politics' of the sort that Mama and Parmar decried, but it is also a more insidious form of ethnocentrism, one which claims, on racial terms, the authority to represent Asianness, but in the process reproduces the representational vocabulary of the colonizer.

Before turning to the four theatre and film scripts that constitute the bulk of my analysis, I want to establish a critical frame by contrasting the rather different identity politics of two well-publicized British-Asian texts, Gurinder Chadha's film *Bend It Like Beckham* (2002) and a novel published a year after *Beckham* was released, Monica Ali's *Brick Lane* (2003). *Bend It Like Beckham* provides a good example of the kind of ethnic self-caricature that I have described above. Jess and Joe may get to kiss at the end of the film, but they break few boundaries, unlike Omar and Johnny in *My Beautiful Laundrette*. Instead, *Bend It Like Beckham* heterosexualizes and domesticates *My Beautiful Laundrette* while adopting its narrative formula: an ambitious young Asian succeeds, with the help of working-class whites, in overcoming the stifling conventions of a traditional family. Kureishi uses this formula to undo all kinds of stereotypes – nothing in *My Beautiful Laundrette* or *My Son the Fanatic* (1998) neatly matches expectation or type, whether it is family life, Thatcherite enterprise culture or sex. In *Bend It Like Beckham*, however, we meet exactly the kind of immigrant family we expect to see in film: a kind and loving but misguided father, a mother who determinedly polices her daughter's marriage prospects by insisting she learn to cook, a transgressive younger generation that cleverly outwits its elders, and a grotesque parade of intrusive aunts and other inane relatives who are always very shocked. Predictably, Jess finds her salvation in football, here represented as the antithesis to life as an Indian wife, and in Joe, who is white (if Irish) and of course a man.

Were it not for its immense popularity in a context in which mainstream films about India or Indians are scarce, *Bend It Like Beckham* would be irrelevant to this discussion. However, coming on the heels of Meera Nair's hit film *Monsoon Wedding* (2001), it attracted unexpectedly large audiences and demonstrated the power of commercial film to invent and define entire cultures and histories. In the process, *Bend It Like Beckham* also demonstrated the success of multinational capitalism in appropriating radical discourses of all sorts, such as those of racial and transnational hybridity – in this case, commodifying both feminism and multiculturalism and cynically deploying their utopian imagery in the global marketplace. Viewers who had never been to Britain, who had never met an Indian family, who didn't quite follow any of the film's accents, who couldn't catch the ethnic issues that complicate the more obvious racial differences, nevertheless understood the film at once. Its plot was comfortingly familiar, a story about tradition-bound immigrants whose daughters yearn to be free by playing ball and marrying for love.

The parodic reappropriation of the icons of cultural authenticity by younger voices can certainly be a radical gesture, simply by parading difference in a community which enforces whiteness. However, it is the success of *Bend It Like Beckham* in positioning *gender* as the litmus test for modernity that is most interesting and relevant here. *Beckham* is particularly comforting in its gender politics, confirming hackneyed conceptions about tradition and modernity while celebrating girl power that never threatens to become feminist and indulging in garden-variety misogyny in its depiction of the mothers and aunts. Above all, it reassures its Western viewers of the desirability of Western values and opportunities *for women*. The film thus offers one version of a narrative of Westernness and modernity in which the assertion of a woman's rights is represented as being incompatible with allegiance to the racial or ancestral past. This formula is of course routinely invoked by ethnic absolutists, religious leaders and other regressive patriarchal forces to condemn feminism as a decadent practice, but it is also an unexpectedly frequent refrain in ostensibly progressive discourses which appeal to feminism in order to extol Western values. In both cases, feminism comes to represent, and to be represented by, the West.

The perceptive conceptual frame of the acclaimed 2003 novel *Brick Lane* offers an interesting contrast to Chadha's film. *Brick Lane* is a story about immigrants written by Monica Ali, a British writer of Bangladeshi/ Muslim and British descent.[15] What happens, Ali asks, when a village girl from Bangladesh – unspoilt, simple, 'the real thing', in the eyes of both her husband and her lover – is tossed into the bewildering post-modern complexity of Bangla London (London's Tower Hamlets)? How is this woman repositioned and reinvented as a global citizen, and does she have any say in her own refashioning? Ali deftly maps the conflicting cultural imperatives that befuddle her protagonist, Nazneen; neither her marriage nor her affair provides her with a road map for determining who or what her post-Bangla self could be or should become.[16] Nor do Nazneen's matriarchs in London, the two elder Bangladeshi women who offer two contrasting models of immigrant womanhood. On the one hand there is Mrs Azad ('azad' means 'free' in Urdu), who appears to Nazneen as a terrifying example of a Westernized, modern British-Asian woman; on the other, we find the woman who stayed true to her roots, the also aptly-named 'Mrs Islam', a deceptively motherly woman who turns out to be the estate's mafia-style usurer and enforcer. Representing the two major types of first-generation women (the assimilationist and the traditionalist), they present Nazneen with grotesque caricatures of the diasporic subject: neither tradition nor modernity as defined by these examples offers Nazneen a way of being anything other than herself. Ali therefore imagines a third way in which Nazneen can meet the demands of time and place: she forms a garment-makers' collective with other dislocated women. Gesturing towards an older generation of the British Left as well as towards British socialist-feminists of the 1960s and 1970s, Ali finally suggests that women (in this case immigrant women) deal with modernity as they have dealt with revolutions of the past, by working, scraping, sewing, designing and mending. In this respect, Nazneen's fate is not significantly different from that of a woman in Bangladesh (the novel charts the parallel adventures of her sister Hasina to underline the point). The novel suggests that the new global citizen is simply a transplanted sister of the millions of women throughout the world who hold homes and communities together.

Ali's conceptualization of the immigrant woman's agency and capacity for self-renewal is significantly at odds with Chadha's crude narrative of generational conflict within British-Asian communities. It is the latter that we most often find in immigrant drama: typically, a tradition-bound family, an abusive husband and a timid (often immigrant but sometimes second-generation) wife silenced by her loyalty to the kinship structures. Rather than examining (as Jackie Kay's play *Chiaroscuro* does) how the claiming of ancestry enables black British women to resist the ways in which their citizenship is underwritten and directed by the state, patriotic allegiance to an imagined ancestral history is seen by works like Chadha's only to discourage women from claiming the rights and privileges they enjoy as inhabitants of a modern state. The relationship between citizenship and ancestry is particularly interesting in its representation of the modernity of British statehood and the pre-modernity of the imagined homeland. What is also intriguing is the way in which the state, usually invisible or implied to be tyrannically patriarchal in radical British theatre,[17] so often appears in such performances not only as the new citizen's legal protector, but also as the force which guarantees her modernity.

To explore how such conflicts have been depicted (and complicated) in theatre and less commercial film, I will examine four scripts by three writers, performed about 15 years apart: Rukhsana Ahmad's *Song for a Sanctuary* (1990), Meera Syal's screenplays *My Sister-Wife* (1992) and *Bhaji on the Beach* (1993), and Gurpreet Kaur Bhatti's *Behzti* (2004). These writers represent three stages in the emergence of British-Asian feminist writers since the mid-1980s. Ahmad came from Karachi (Pakistan) to Britain in 1973 in her mid-twenties, after her marriage, and was by the mid-1980s actively involved in British-Asian feminist writing and theatre circles, including the Asian Women Writers' Collective. Syal, now an established and widely acclaimed second-generation British-Asian writer, was awarded the Betty Trask award in 1996, an MBE in 1997, a Commission for Racial Equality media award in 2000, and a number of other awards besides. Bhatti is a relatively new voice in British theatre who achieved an unfortunate notoriety for her second play *Behzti*, which was shut down in December 2004 after it provoked a riot in Birmingham; she was awarded the 2005 Susan Smith Blackburn Prize shortly after.

Rukhsana Ahmad, *Song for a Sanctuary* (1990)

Song for a Sanctuary was the first play produced by Kali theatre company, which Rukhsana Ahmad co-founded with Rita Wolf in 1989. A modest, thoughtful and tidily constructed work in the style of British-Brechtian educational theatre, Ahmad's play offers a nuanced exploration of negotiations involving generation, ethnicity, class and ideology. These are presented in two parallel sets of conflicts, on the one hand between two women of Indian descent and on the other between two generations of feminists. Set in a women's shelter, the play focuses mainly on the experiences of a Sikh woman named Rajinder, who, accompanied by her three children, has left her abusive husband Pradeep. As a middle-class woman of Punjabi origin (as a Punjabi Sikh she is almost certainly from India), she is seen by herself and others as an 'atypical' resident, and from the outset irritates Kamla, one of the younger workers at the shelter, who is a second-generation woman of Indo-Caribbean descent. Rajinder's class standing, make-up, demeanour and unwillingness to divulge her reasons for leaving home make Kamla suspect that Rajinder doesn't have a legitimate reason for being at the shelter: 'she's not the kind who needs help', Kamla tells her older co-worker Eileen. Eileen, however, urges her to look more deeply, articulating the core hypothesis of the play:

> How many times 've you said yourself that class divides women, amongst other things... Try to find the common ground, there must be some, somewhere within her, there must be the woman who needs help.[18]

Unwilling to take Eileen's advice, Kamla harbours a complex hostility towards Rajinder throughout the play, equating Rajinder's cultural and religious ties with a deep-rooted social conservatism and aversion to change. Kamla expresses surprise that Rajinder prays ('In this day and age?'), particularly as she seems to do so in the middle of the night: 'It's not adult behaviour', she argues, 'We really do need to do some consciousness-raising work here.' In response to Eileen's gentle suggestion that praying is much the same as hoping, Kamla rages, 'She's just like my mother. You won't destroy her faith. It would take a bloody miracle to do that.' Warning that Rajinder's reluctance to reveal the truth of her domestic circumstances signals that she is one of the women who 'are collaborators by nature', Kamla argues that Rajinder is wholly 'un-critical of the outside world' (Ahmad, *Song*, pp.168–9).

Although Eileen repeatedly tempers Kamla's uncompromising judgements, the play supports Kamla's reading in key ways: Rajinder is indeed a snob who is stubbornly unwilling to attend group therapy sessions with other women, not wanting to be treated like 'illiterate working class women'; she conservatively monitors her teenage daughter's clothing and entertainment, insisting, 'We're different from all these people here' (Ahmad, *Song*, pp.174–6). We gradually learn that she has more in common with the other women than she would like to admit: she has been beaten by Pradeep, forced to drink, and made to participate in his sexual fantasies in deeply humiliating ways.

At the same time, the audience is occasionally allowed to see that Rajinder has privately made a significant psychological break from her husband, and that her conservatism belies the depth of her commitment never to return to Pradeep. She has been to a shelter before only to return to her husband, but we are asked to believe her when she vows that she has 'planned it all carefully this time' (Ahmad, *Song*, p.161). Ahmad sensitively charts the profound familial alienation that Rajinder has experienced in leaving Pradeep: in a flashback, we see the bitter conversation she has had with her sister Amrit, who berates her for bringing a killing burden of 'shame' and 'disgrace' to their elderly parents. We discover that Rajinder has often begged to be allowed to return to her maternal home, only to be rejected by a family that prefers an honourable death to a dishonourable divorce.

But Rajinder is herself too tied to notions of family honour and privacy to be able to discuss her life with social services representatives such as the housing officer, or for that matter with other women at the shelter.[19] This silence, the play emphasizes (echoing Kamla), endangers Rajinder, her daughter and the shelter itself. When Rajinder discovers that another resident at the shelter, an occasional prostitute, has brought a male client to her room, she is horrified and decides to risk returning to her husband's home. There she discovers what the audience has known for some time, that her daughter Savita has been sexually abused by Pradeep. However, while Savita has feared that her mother would not believe her, or, worse, blame her, Rajinder immediately takes action by returning to the shelter. She nevertheless remains entirely unwilling to report Pradeep for abusing their child, as Kamla pushes her to do: Kamla argues that taking the matter to the courts will solve the entire problem by insuring housing and counselling for Savita and prosecution for Pradeep, while Rajinder prefers to help Savita forget the whole thing, convinced that the publicity of a trial will only compromise Savita's future, and ruin her prospects for marriage and communal acceptance.

Eileen intervenes, promising Rajinder that things will work out. In the contrast between Eileen's method of dealing with Rajinder and Kamla's, Ahmad explores the critical differences between two generations of feminists: an older generation with a strong conviction that key common experiences unite all women, and a younger generation that is concerned with differences of race, ethnicity and class on the one hand (and therefore with the need to re-educate women who have internalized sexist attitudes), and on the other hand with the importance of a legal and governmental infrastructure for dealing with private lives. For the most part, the play privileges Eileen's compassion and insight while suggesting that Kamla is, as Griffin puts it, 'rule-bound, authoritarian, overly assertive, and insensitive to Rajinder's cultural needs' (*Contemporary*

Black and Asian, p.156). However, only two scenes after Rajinder refuses Kamla's advice and cries on Eileen's shoulder, Pradeep infiltrates the shelter and stabs Rajinder to death. In the final, ironic statement of the play, another resident assures the sobbing Savita, 'Someone will be here to help us soon, I'm sure' (Ahmad, *Song*, p.186). By the end, then, the play recognizes the importance of Kamla's attention to the 'rules', to law, and suggests that the state must develop stronger support systems for immigrants who need to escape abusive homes.

The play ultimately suggests that neither the home nor traditional kinship ties can help immigrant women protect themselves; rather, they must turn to the security provided by the state and civic authorities to break free. Ahmad proposes that rigid and uncompromising ties of religion, family and community can be hypocritical and destructive, but she stops short of suggesting that immigrant women should reject ethnic affiliations altogether. The problem, the play suggests, is not an issue of ethnicity so much as generation. Significantly, the modern woman in *Song for a Sanctuary* is not white; indeed, Eileen's liberal acceptance of Rajinder's need to work within her cultural norms is seen to be part of the problem. Rather, the new citizen is Kamla, who has the clearest and most cogent feminist analysis, even if she does lack people skills. It is interesting that the play positions Kamla as someone who is beyond ethnicity. The first scene of the play ends with Kamla's frustrated monologue about her parents' failed attempts to Indianize her – 'Language classes, music lessons, dance lessons, they tried it all . . . it was no use to me. Who cares for all that crap anyways?' (Ahmad, *Song*, p.164). Her encounter with Rajinder provokes memories that are at once tender and unhappy; she remembers lessons on how to fold a shawl properly, sings snatches of an old romantic song from a Hindi movie, and stiffly tries some half-remembered steps of Kathak. The play thus suggests that Kamla may be transferring on to Rajinder some lingering resentments about her own upbringing, but also makes clear that it is only by refusing those ties that she has been able to blossom as a feminist. This sort of change, Ahmad reminds us, is neither easy nor quick for a woman of Rajinder's background – but she warns us also that Rajinder does not have the luxury of time.

Meera Syal, *My Sister-Wife* (1992) and *Bhaji on the Beach* (1993)

In two particularly nuanced and provocative feminist works by Meera Syal, *My Sister-Wife* (1992) and *Bhaji on the Beach* (1993), generational conflicts similarly structure a dialectic of tradition and modernity that is explained in gendered terms. *Bhaji on the Beach* was directed by Gurinder Chadha (the director of *Bend It Like Beckham*) and received a positive if limited reception. *My Sister-Wife* was performed as a three-part serial on BBC; I discuss it here not because it is necessarily representative of Syal's work, but because of its particularly fascinating representation of the cultural politics of British Muslims.[20]

In both works, Syal suggests that coercive religious and kinship bonds keep second-generation black British women trapped in self-destructive relationships with their families, communities and histories. On one side we find the forces which compel women to consider breaking away: most importantly the physical and psychic violence of patriarchy, and, less visibly but crucially, the implicit promises of citizenship and modernity – the state's guarantee of individual rights related to sexual practice and marriage. On the other side lie the constant threat of assimilation into 'Western decadence' and, perhaps most importantly, the seductive security of cultural coherence. These fears and desires, Syal argues, lead younger women to silence themselves, subordinating their own needs and rights.

My Sister-Wife

My Sister-Wife imagines a situation in which Farah, a young British-Asian woman of Muslim, Pakistani origin, born and raised in Britain, voluntarily agrees to live bigamously.[21] The script establishes at the outset that Farah is educated, independent and well employed, and that she and Asif are in love, not traditionally arranged to be married by their immigrant parents. Syal thus begins by ruling out the obvious socioeconomic factors which might ordinarily explain the cultural contexts of polygamy, focusing instead on the complex, contradictory psychological factors which underpin second-generation ventures into what Syal depicts as the perilous, alien, liminality of transnational culture.

Farah is not afraid of living between cultures; on the contrary, she begins with the assumption that she shares with Asif not only a compatible multicultural heritage ('he understands all of me...both sides') but also a sense of what is rational and legal, qualities which the play equates with the West and modernity. She therefore believes his explanation that he married his first wife Maryam only out of a sense of duty to his family, which obligation also requires that he continue to live with her and their two daughters (and his mother). Farah's own mother Mumtaz endorses this explanation and encourages her to marry Asif in spite of Maryam: 'He'd be no man in my eyes if he could forget his responsibilities towards her' (Syal, *Sister-Wife*, p.113). The British state does not recognize Asif's first marriage as it has never been registered in the United Kingdom, which gives Farah some confidence in her legal standing. Farah also believes that Asif, as a modern man, will eventually leave Maryam and that they will move into a separate flat.

However, although Farah enters a polygamous arrangement apparently without a full understanding of its implications for her life with Asif, she gradually finds herself both accepting and resisting what Syal represents as a traditional Pakistani family structure. When (soon after their marriage) she begins to suspect that Asif is sleeping with Maryam again, Farah is pulled in several directions. Her 'Western' side, predictably, reminds her that she is 'intelligent' and solvent, and urges her to leave Asif. Then a submerged longing to claim an 'Eastern' home and heritage is deeply intensified by her mother's death and her father's subsequent departure for retirement in Pakistan. These events lead Farah to decide unexpectedly that she will try to live with Maryam in the house after all: 'We do things differently', she resolves in a sudden reclaiming of her ancestral ties, 'Family, duty, responsibility...I've thrown so much of my past away without trying to understand it' (Syal, *Sister-Wife*, p.120).

Thus although she continues to see Maryam as a sexual rival, Farah tries to reconcile East and West imaginatively, seeking an interpretive matrix which will provide her with the means to reconcile her conflicted cultural imperatives. Importantly, it is gender that provides Farah with the means to imagine her own diaspora space. In a dream vision, she reconstitutes the harem as a community of women, as a 'paradise'. Syal's directions read:

> We follow Farah through a maze of marble corridors, exquisitely carved pillars, silk wall hangings, Mughal miniatures. Somewhere there is sitar music playing.... A group of four women, ranging in age from fifty to sixteen, all recline on silk cushions, drinking tea and feeding each other with sweet-meats whilst young children play around their feet.... Sunlight streams through the carved wooden shutters at the windows throwing moving spirals of pattern over their faces. Utter peace. The women smile at each other, beautiful in their contentment.
>
> (Syal, *Sister-Wife*, pp.127–8)

Derived from her mother's sentimental recollections of the 'perfect' life of her father's polygamous cousins in Lahore, Farah's vision reproduces exactly her parents' immigrant idealization of the roots from which they imagine they have been dislocated. More interesting still is Syal's evocation of the languages of colonial desire and nostalgia (this could be the setting for an Ivory-Merchant film), an association which allows viewers to see what Farah has forgotten, that her craving for ethnic authenticity and acceptance is itself a function of her postcolonial Britishness.

For a brief time, this reinterpretation allows Farah to experience what she misrecognizes as the traditional securities of the harem: when Asif leaves for a trip, Maryam and Farah forge an attempt at sisterhood, sharing recipes and make-up tips. But it soon becomes clear that the new relationship is simply a new medium for their sexual rivalry: by the time Asif returns, each woman has transformed herself into the other in an effort to win his love: Maryam wears a lounge suit and is impeccably made up, while Farah, in a salwar kameez, has grown her hair long and ties it back. But Farah's attempts to be 'Pakistani enough' fail, she is told, because she continues to want to 'break up this family' by demanding a private relationship with Asif. She is furious to find herself cast as 'a bastard to my own people' (Syal, *Sister-Wife*, p.140).

After Farah is incapacitated by a difficult pregnancy and miscarriage, she gradually loses her ability to think of breaking away. Maryam takes over Farah's job and gradually replaces her as the modern wife (a performance which, absurdly, includes the acts of talking back to Asif and his mother, wearing Western clothes and finding a job). Farah becomes a 'mad-woman in the attic', as Asif says; she voyeuristically keeps tabs on Maryam and Asif's sexual habits, takes over the housekeeping, and wears traditional Pakistani clothes. Eventually, convinced Maryam is trying to kill her, she attempts to poison Maryam and ends up killing Asif instead. The film ends with an ironic reinvention of the harem of Farah's vision: Maryam sits with Farah on a bed and rocks her consolingly.

As May Joseph notes, *My Sister-Wife* 'shreds the illusion that education and modernization can buy middle-class women enough consciousness to opt out of the patriarchal circuits of exchange and subjection'.[22] But Syal's feminist critique is predicated on a very simple binary in which patriarchal privilege and tyranny are functions of 'pre-modern' Muslim/ Pakistani culture (the two are unproblematically conflated), while the promise of feminist self-realization and consciousness lie exclusively in the women's ability and willingness not only to claim the rights and protections available to them as British citizens but also to abandon altogether the ties of tradition and ethnicity. As Griffin puts it:

> the females try to negotiate, for the most part unsuccessfully, between the traditions with which they have been brought up and

> which insist on the primacy of the family, the community, *izzat* [honour] and *sharam* [shame] as the guiding values for one's actions, and a western position that favours individualism, self-determination, equality between the sexes, and the economic independence of women.
>
> (Griffin, *Contemporary Black and Asian*, p.168)[23]

This dichotomy leaves both categories, 'Pakistani/Muslim' and 'Western', wholly uncontested and monolithic.

Although *My Sister-Wife* is quite profoundly reductive and ethnocentric in its depiction of cultural difference, it is an important statement about the relationship between the new citizen and her past. Farah's ill-conceived acceptance of a polygamous marriage is also a metaphor for all second-generation ventures into the past, and in this (single) way *My Sister-Wife* resembles Kureishi's *My Son the Fanatic,* which similarly warns of the dangers of uncritical engagement with incompletely understood ethnic histories. The complexity and insight of *My Sister-Wife* lie in Syal's sympathetic if cautionary portrayal of Farah's longing for an ethnic home to be nostalgic about; Syal suggests that the inability to relinquish a sentimental tie to an imagined history endangers rather than empowers women.

Bhaji on the Beach

Although she advances a similar argument in her film *Bhaji on the Beach,* Syal is more optimistic about the possibilities of sisterhood and much more nuanced about her sense of the interplay of ethnicity and gender in the fashioning of modernity.[24] In the film, a group of Indian women of different ages take a day trip to the seaside resort of Blackpool; the trip is organized by Simi, the head of Saheli, a centre for South Asian women ('saheli' means 'girlfriend'). The older (immigrant) women police second-generation gestures towards a series of what Syal suggests are acts of emergent modernity: a young mother, Ginder, decides to leave her abusive husband; Hashida, a medical student with an unblemished social profile, is having a secret affair with a black British lover and discovers they are pregnant; more trivially, the younger women sport trendy clothes and publicly display an interest in young men. The older women's criticisms are finally overcome when, in the familiar pattern of 1970s consciousness-raising, the abusive behaviour of Ginder's husband becomes public and the women rally in support of his victim.

Although Syal is mainly concerned with the ideological factors which try to repress the modernity of the younger women (such as Ginder's brief decision to return to her husband), she also pays close attention to the self-questioning which is gradually undertaken by one of the older women, Asha, whose melodramatic movie-inspired hallucinations punctuate the film both as comic interludes and moments of significant cognitive crisis. In a subtle portrayal of Asha's smothered desires and rich fantasy life, Syal shows Asha carefully scrutinizing the motivations which led her to choose family and duty over self (that is, tradition over modernity), and Asha's own ambivalent sense of the consequences of her choices.[25] She comes to realize that what she has valued most about herself are precisely those elements of imagined Indian womanhood traditionally sanctified in popular film and religious myths: elegance under duress, discretion, an imperturbable spiritual centeredness. In one of the film's most interesting and funny scenes, Asha imagines herself cavorting, Bombay-film-style, around the trees at Brighton Pavilion with her new friend Ambrose, an old-fashioned gallant entranced by Asha's palpable old-world authenticity – she imagines him in brown-face, and her vision is abruptly shattered as the rain washes away his colour to reveal his white skin. In this amusing and insightful moment, Syal reminds us that the iconic Indian womanhood celebrated in Indian films is also a cherished image of old British India, eroticized in countless narratives about the beautiful dangers of miscegenation.

Asha, whose name means 'hope' or 'expectation', eventually reaffirms her connection to her roots, rediscovering its old pleasures and securities. However, her visionary inner journeys allow her to come to a better understanding of the gulf which divides her cultural centre from the new British Asia, and this awareness prompts her to reject her previously uncritical endorsement of traditional kinship bonds. In the early parts of the film, she had encouraged Ginder to return to her abusive husband Ranjit; by the end of the film, she sees the truth and intervenes when Ranjit tries to hit Ginder, slapping him.

In both films, Syal's representation of the past, of Pakistan or India *as* the past, is at once subtle and troubling, and reproduces the pattern we have seen in *Bend It Like Beckham*. However, *Bhaji on the Beach* also provides a comic snap-shot of a real 'diaspora space', a Blackpool Beach carnival in which both ethnicity and gender are brought into play in such a way that none of the characters – white, Asian, black, male or female – is allowed to emerge with identity securely intact. The quiet, backstage agent of this transformation is Simi, a second-generation feminist, but Syal insists through her depiction of Asha that the first generation must participate fully in the process of change and modernization. In this respect the film echoes *Song for a Sanctuary*, which similarly suggests that significant changes in the lives of diasporic South Asian women must involve a negotiated engagement of ideological and identity-related differences between first- and second-generation women.

Gurpreet Kaur Bhatti, *Behzti* (2004)

The conflict between tradition and modernity also underpins Gurpreet Kaur Bhatti's controversial play *Behzti*, which was forced to close during its opening run in December 2004 at the Birmingham Repertory Theatre, following the riotous conclusion of an initially peaceful protest by hundreds of members of Birmingham's Sikh community.[26] Religious authorities consulted by the theatre had requested that the play's climactic scenes, involving rape and murder, be rewritten so that the events would not take place in a gurdwara (a Sikh temple); they suggested a pub or nightclub instead. The protest was organized after Bhatti expressed an unwillingness to change the setting. (As is quite typical in controversies involving theatre alleged to be indecent or offensive, most of the protestors had not seen or read the play.)

The controversy focused, perhaps inevitably, on the insult implicit in the play's setting – the play's quite unambiguous suggestion that the gurdwara has lost its moral authority, that beneath its respectable façade lurk deep-seated fears, shameful desires, and a long history of duplicity and viciousness. '[S]ometimes I feel imprisoned by the mythology of the Sikh diaspora', writes Bhatti in her foreword to the play:

> Clearly the fallibility of human nature means that the simple Sikh principles of equality, compassion, and modesty are sometimes discarded in favour of outward appearance, wealth, and the quest for power. I feel that distortion in practice must be confronted and our great ideals be restored. Moreover, only by challenging ideas of correct and incorrect behaviour can institutionalised hypocrisy be broken down.
>
> (Bhatti, *Behzti*, p.17)

The play focuses on Balbir Kaur, a foul-mouthed, disabled Sikh widow in her mid-fifties, and her efforts to find a husband for her 'faithful', innocent and 'ungainly' daughter Min (Maninder), who is 33. Accompanied by Balbir's black Social Services caseworker, Elvis, they visit the gurdwara on Guru Nanak's birthday (the holiest day of the Sikh calendar), in search of a matrimonial list reputedly kept by Mr Sandhu, a pillar of the community and senior officer at the gurdwara. In a Gothic climax, Min is raped by Mr Sandhu, who, it transpires, has not only raped many other girls and boys in the inner sanctum, with the evident co-operation of the community, but desires Min in particular because her father Tej (an alcoholic who has committed suicide some years ago) had once been his lover. After these shocking secrets are revealed, Min's mother Balbir finally comes to her defence by killing the rapist, and Min discovers a tender and true love with Elvis, who promises to teach her to dance.

Behzti, which means 'dishonour', does not offer a reasoned argument about religious hypocrisy or community betrayal. Rather, the play blazes through home and sanctuary, demolishing myths at each step; it argues quite clearly that the institution no longer has the capacity to reform itself, and that the gurdwara can no longer house the faith. Interestingly, while Balbir and Min are seen to have been crippled and disfigured by poverty and the community's neglect, Bhatti suggests that they have also enjoyed a kind of rambunctious, unconventional camaraderie. Their isolation from the community has allowed them to develop an edgy iconoclasm and unfeminine boisterousness: Balbir gleefully and sacrilegiously eats beef on her way to the gurdwara, while Min responds ecstatically if atypically to the stirring sounds of Sikh prayers by dancing to Michael Jackson's 'Billie Jean'. In one scene Min sellotapes her mother's mouth and ties her wrists so she can speak to her (and astonishingly it works). Most importantly, Bhatti suggests that their removal from gurdwara culture (and more generally from their community) has allowed Min's religious faith to remain childlike and honest, a true faith.

The crisis of the play is precipitated by Balbir's need to reconnect with the community, a decision the faithful Min naively applauds in spite of her shyness, as well as by Balbir's insistence that it is necessary to 'settle' Min, that is, to arrange a marriage, as she is evidently not interested in actively pursuing a mate for herself. The play gently condemns Balbir for her crassly materialistic fantasies – 'a rich, handsome, successful fiance', whose 'big property' will include private quarters for Balbir – but also shows some sympathy for her desire for something better than the dishonourable 'shitter hole' Min's father has left them in. When Min asks why they can't be 'happy as we are', Balbir furiously asks how they are to do so:

> Eating plastic frozen school dinners, waiting for you to wipe my arse, being wheeled about like shopping in a supermarket trolley? You think it is pleasant watching a fat virgin become infertile? I want to live a life that is something. I want to be seen and noticed and invited by people. I want anything... that is not this.
>
> (Bhatti, *Behzti*, p.46)

Ironically, of course, it is Balbir's decision to reassimilate, to seek community sanction and honour for her daughter through marriage and religion, that eventually shatters Min's innocence and almost destroys her. Tradition, whether religious or communal, is seen by Bhatti as a coercive, destructive force in the lives of first- as well as second-generation immigrants, particularly women, while the chaotic dance of modern multicultural Britain offers possibilities for real happiness and love, at least for Min and Elvis.

Conclusion

The plays I have considered are, on the whole, pessimistic about the promise of transnational or transcultural alliance, at the same time that they are written by some of the women best positioned to imagine feminism in global terms. However seductive the allure of tradition and ancestry, they suggest, feminists must remember that the past confuses, betrays and ultimately destroys women, and that feminist self-realization and fulfilment are incompatible with ethnic ties. In other words, the plays imply that diasporic feminists should not be primarily concerned about re-theorizing the nation or transforming nation-states into diaspora spaces, and should focus instead on carving spaces within the British state where first- and, particularly, second-generation feminists can safely achieve a more fully realized citizenship. Shelters are an important example of such spaces, and 'social services' are seen as midwives in the process by which women emerge from the enslaving confines of tradition into the liberating light of British modernity. The plays acknowledge that the state's provisions are imperfect, but nevertheless insist that such provisions are important safety nets for first- and second-generation women whose relationship to their families or communities confines or endangers them.

Some of the texts I have considered are 'youth stories', and I acknowledge that my readings, with the eyes of a first-generation Indian, may seem jaundiced.[27] However annoying the performance of Asianness in *Bend It Like Beckham* might seem, it's also important to remember that, for second-generation writers of all backgrounds, multiculturalism often involves 'talking in tongues', to use the title of Winsome Pinnock's play.[28] Postmodernist repudiations of absolutist notions of ethnic purity or cultural authenticity go hand-in-hand with the side-shows of institutionalized multiculturalism, which include not only the parodic elements I have criticized but also 'celebrations' of racial or ethnic identity – parades, festivals, mass religious events, fashion shows – that define affiliations and loyalties along equally predictable lines.

But such performances do not provide feminists with a useful way of conceptualizing a transnational practice. Perhaps Monica Ali's novel is right in its suggestion that transnational feminism can be imagined only through a recognition of women's work and agency. One of the first spectacles of the twenty-first century was that of Amina Lawal, a young Nigerian woman who was sentenced in March 2002 to death by stoning for the crime of extra-marital sex. In pictures distributed widely through television and the internet, Lawal stood poignantly holding the main evidence of her crime, an infant daughter, to her cheek: the very picture of the victim of a world rapidly going mad in the hands of resurgent fundamentalisms, and a stirring example (underscored by the excesses of the Taliban) of the need for an active, international, even interventionist feminist movement that would transcend the mushy rhetoric of cultural relativism to affirm women's rights in unapologetic, unambiguous terms.

Also widely distributed were pictures of Lawal's acquittal in September 2003, in a small sharia court of appeal. Lined up behind her, unremarked in news reports, was her team of lawyers: a group of Nigerian women in traditional dress with headscarves. None of the reports of Lawal's sentence mentioned the lawyers who had argued her case, or Baobab, the Nigerian women's rights organization that had organized Lawal's defence as well as interventions in similar cases. An obvious first step in fashioning a transnational feminism would be to recognize the modernity of the other women's intervention – to reframe Lawal's plight in terms of Baobab's effort on her behalf. As Ayesha Imam of Baobab put it, a transnational intervention must 'pay attention to people who are directly involved in the situation on the ground – and the wishes of the people whose rights we are trying to defend'.[29] For all feminists, this entails acknowledging feminist work internationally to be real, legitimate and as modern as any feminism in the West.

Notes

1 Gabriele Griffin, *Contemporary Black and Asian Women Playwrights in Britain* (Cambridge: Cambridge University Press, 2003).
2 Avtar Brah, *Cartographies of Diaspora: Contesting Identities* (New York: Routledge, 1996), pp.208–9.
3 Eric J. Hobsbawm, 'Ethnicity and Nationalism in Europe Today', in Gopal Balakrishnan, ed., *Mapping the Nation* (London and New York: Verso, 1996), pp.255–66, p.255.
4 Stuart Hall, 'New Ethnicities', in Houston A. Baker, Jr, Manthia Diawara and Ruth Lindenborg, eds, *Black Cultural Studies: A Reader* (Chicago and London: University of Chicago Press, 1996), pp.163–172, pp.168–9.
5 Nira Yuval-Davis, 'Women, citizenship, and difference', *Feminist Review*, 57 (Autumn 1997): 4–27, p.8.

6 Pratibha Parmar, 'Other Kinds of Dreams' [1989], in Heidi Safia Mirza, ed., *Black British Feminism: A Reader* (London and New York: Routledge, 1997), p.69.
7 Amina Mama, 'Black Women and the British State: Race, Class and Gender Analysis for the 1990s', in P. Braham, A. Rattansi and R. Skellington, eds, *Racism and Anti-Racism: Inequalities, Opportunities and Policies* (London: Sage Publications, 1992), pp.79–101, p.97. Mama attributes this development to the state's response to Brixton, which was to devise and fund a variety of 'ethnic' social and cultural programmes, one consequence of which was the emergence of a wide variety of black and Asian voices in the theatre.
8 Mary Karen Dahl, 'Postcolonial British Theatre: Black Voices at the Center', in J. Ellen Gainor, ed., *Imperialism and Theatre: Essays on World Theatre, Drama, and Performance* (London and New York: Routledge, 1995), pp.38–55, p.46. See Dahl's essay for a thoughtful discussion of the relevant terminology.

   The categories of 'immigrant' and 'second generation' as I use them are not sociologically rigorous, and they do of course intersect to a significant extent. Is Salman Rushdie a migrant or a new citizen? He was born in Bombay, but was sent to school in England before he was 15; I would argue that while he represents both to an extent, his voice is more second generation than first.
9 Paul Gilroy, '"To Be Real": The Dissident Forms of Black Expressive Culture', in Catherine Ugwu, ed., *Let's Get It On: The Politics of Black Performance* (London: Institute of Contemporary Arts, 1995), pp.12–33, p.15.
10 This is a synopsis of the discussion in my essay 'Small Island People: Black British Women and the Performance of Retrieval', in Elaine Aston and Janelle Reinelt, eds, *Cambridge Companion to Modern British Women Playwrights* (Cambridge: Cambridge University Press, 2000), pp.217–34.
11 Kobena Mercer, *Welcome to the Jungle: New Positions in Black Cultural Studies* (New York and London: Routledge, 1994), pp.4–5.
12 Interestingly, Afro-Caribbean theatre practitioners of the 1950s and 1960s gravitated towards television (and later film) by the 1970s, and a similar pattern is noticeable among writers of South Asian descent who came of age during the 1980s. The establishment of Channel 4 in 1982 had a particularly important impact upon the emergence of South Asian voices in British film – it was there that the popular comedy *Goodness Gracious Me* got its start, for example.
13 Ayub Khan-Din, *East Is East* (London: Nick Hern Books, 1997), p.57.
14 Samina Zahir, 'Goodness Gracious Me', in Alison Donnell, ed., *Companion to Contemporary Black British Culture* (London and New York: Routledge, 2002), p.128.
15 Monica Ali, *Brick Lane* (London: Doubleday, 2003).
16 Her husband and lover seem at first to represent the familiar binary of tradition and modernity, but this paradigm is revealed to be significantly more confusing and complex than one might imagine after watching *Bend It Like Beckham*. In a refreshing change, no starry happy ending follows neatly upon Nazneen's affair with Karim: she is not modernized, liberated or indeed changed very much at all after her affair with him. The men's destinies are much more clearly defined. Nazneen's Mr. Biswas-like husband Chanu, 20 years her elder, eventually chooses to return to Bangladesh, while

her lover Karim, a young Bangla-Briton, dons a kaftan and leaves to join the *jihad*. Because Karim is initially positioned as the new citizen – the one truly capable of uniting the contradictory strains of Bangla ethnic identity, British citizenship and transnational Islam – Ali reminds us through his example of the dangers of community affiliations that are too narrowly conceived.

17 This is particularly true in British socialist theatre of the 1970s and 1980s, but I am also thinking of such feminists plays as Caryl Churchill's *Cloud Nine* (1979) and *Top Girls* (1982), and Pam Gems's *Queen Christina* (1977).

18 Rukhsana Ahmad, *Song for a Sanctuary*, in K. George, ed., *Six Plays by Black and Asian Women Writers* (London: Aurora Metro Press, 1993), pp.159–86, pp.163–4.

19 See Gabriele Griffin's discussion of Rajinder's 'strong sense of *izzat* (honour) and *sharam* (shame)', *Contemporary Black and Asian*, pp.151–2.

20 It is useful to note that Syal is of Punjabi, Hindu and Sikh background (not Muslim); Rukhsana Ahmad, who wrote about a Sikh family in *Song for a Sanctuary*, is of Muslim background.

21 Meera Syal, *My Sister-Wife*, in Kadija George, ed., *Six Plays by Black and Women Writers* (London: Aurora Metro Press, 1993), pp.111–58.

22 May Joseph, 'Borders Outside the State: Black British Women and the Limits of Citizenship', in Peggy Phelan, and Jill Lane, eds, *Ends of Performance* (New York: New York University Press: 1998), pp.197–213, p.204.

23 'Izzat' in this context means respectability or honour; 'sharam' means shame.

24 *Bhaji on the Beach*, directed by Gurinder Chadha, First Look, 1993.

25 See Jigna Desai's useful discussion of Asha's 'dis-Orientation', in *Beyond Bollywood: The Cultural Politics of South Asian Diasporic Film* (London and New York: Routledge, 2004), pp.144–6.

26 Gurpreet Kaur Bhatti, *Behzti (Dishonour)* (London: Oberon Books, 2004).

27 I should also note that I reside in the United States, where the South Asian presence is small compared to Britain; the only 'Indian' many Americans had seen before *Beckham* was Apu in *The Simpsons*.

28 Winsome Pinnock, *Talking in Tongues*, in Yvonne Brewster, ed., *Black Plays: Three* (London: Methuen, 1987), pp.171–227. Pinnock's play explores issues of race, sexuality and loyalty in cross-cultural perspective, moving from London to Jamaica. Pinnock writes that she has 'always loved the biblical story of talking in tongues':

> The idea that the whole world once shared the same language appeals to a certain sentimental idealism. In that story the separation of people is caused by a fall from grace, the ultimate punishment an inability to understand one another.
>
> (p.226)

29 In 2002, Baobab had issued a strong statement against the signature campaigns organized in support of Lawal, deciding 'that [it] had to put out an international appeal trying to clarify the situation and asking people not ... to participate in international protest campaigns' (Ayesha Imam, quoted in 'Amina Lawal Campaign "Unhelpful"', *BBC News* [UK edition] Tuesday, 13 May 2003). See http://news.bbc.co.uk/1/hi/world/africa/3024563.stm. See also the Baobab website at http://www.baobabwomen.org/index.html.

4
Curious Feminists

Leslie Hill and Helen Paris

If someone asks me if I'm a feminist, I say yes without hesitation. I've always said yes and I always will say yes in answer to that question. So that's clear. If, on the other hand, someone asked me if our company, Curious, was a feminist company then I'd have to pause for a moment and think about it because although the directors, Helen and I, are both feminists, I don't really think of the company as a feminist company. Is this a contradiction? I suppose I don't think of Curious as a specifically feminist company because our work doesn't deal primarily with women's issues from project to project, so it wouldn't seem quite the right label for the company. As Curious we have a feminist outlook, but not an exclusively feminist agenda in terms of the work that we make. I am a little uneasy, however, about saying that Curious isn't really a feminist company, as I don't want this to be taken as an anti-feminist or 'postfeminist' stance because I regard anti-feminism as insane, and 'postfeminism' (the notion that we have transcended the need for feminism) as totally out of touch with issues on the ground for women around the world today. I think I make a distinction between my personal feminism and the nature of my company because, as an individual, I call myself a feminist because it is a belief I hold and share with other feminists whether I back this belief up with lots of actions or no actions – whereas the character of the company, to my mind, is located pretty exclusively in action. That is to say that, since Curious isn't a person, Curious can't hold beliefs, so the character of the company is located in the body of work that we make rather than in the views that we hold as writers/performers, and if our primary engagement is not with women's issues, then the company doesn't really seem to be 'feminist' to me in the sense in which many female companies have been feminist.

The strange thing for me is that I hear a lot of female theatre students state for the record to their class mates that they are NOT feminists, and oddly enough they seem to do this particularly if they have formed an all-female group of performers, or if they are making a piece about what could be considered a women's issue. There seems to be some sort of automatic defence mechanism for younger women around saying that they aren't feminists, which is sad, I think. I've never been a big fan of essays that open with a definition from the dictionary, but sometimes it is undeniably helpful to go back to square one and ask, what does this word actually mean? Nowhere is this truer than in the case of the word feminism, which has always been surrounded by angst.

> **Fem•i•nism** *n*: a belief in the need to secure or a commitment to securing, rights and opportunities for women equal to those of men; the movement committed to securing and defending equal rights and opportunities for women equal to those of men.[1]

A great word. A wholesome and worthy definition. How, then, can this be a word that strikes such terror in the hearts of young women today, so hasty to have it known that they are NOT feminists? Why not? Aren't we all feminists? Don't we all generally believe that women should have equal rights and opportunities in our society? How then did 'feminism' come to be such a maligned and mistreated word? Who convinced women to treat the term feminism with such contempt? What do people mean when they assert, 'I'm not a feminist'? Do they really mean to say, 'Well, I don't actually believe that women should have equal rights and opportunities. In fact, I believe that women should have less legal representation, less opportunity, less education and receive lower wages for the same work.' Is that what they mean? Because that's the logical conclusion of saying, 'I'm not a feminist'. I mean, how many of us really think that women's rights is a gut-wrenchingly controversial topic? That women shouldn't have equal rights under law in our society? Not that many of us, I suspect, and that may be our double-sided problem:

1 we kid ourselves that women are already equal in our society (which the most cursory glance at any number of statistics will flatly disprove with the finality of death by guillotine)
2 we think standing up for women's rights is somehow embarrassing and/or not in our personal self-interest.

So looking around at the denial, by younger women in particular, of any vested interest in feminism, the future of feminism seems a little bleak to say the least. And look at me – a self-declared feminist and yet I'm saying that my company, run by four women with two women as its Directors, isn't really a feminist company. It's all very confusing in terms of trying to predict the future of feminism in theatre. And yet... I think feminism, the true definition of feminism, still thrives in the hearts of most women and men, but that our relationship to content and composition as feminist theatre-makers is changing all the time. Just because I don't think that Curious is a specifically feminist company doesn't mean that I don't think we engage with feminist issues, just as many other theatre-makers and performance artists do. So what about the future of feminism in theatre and live art?

I reckon that the future of feminism might lie in mixing and mingling. That is to say that in an era where many women do not feel an urgency about feminist issues, feminist issues will probably mingle in art works as less explicit, more multi-layered sorts of inquiry into cultures and contexts. So as our contribution to thinking about feminist futures, Helen and I wanted to offer a couple of examples from the work of Curious of overt and embedded engagement with feminist themes and issues. We'll start with a couple of pieces where the connection to feminism is overt albeit layered with other issues, *Three Semi-Automatics Just for Fun* (1997) and the *Guerrilla Performance Locator* (2003) and then move to a case study of a piece where the relationship is more embedded, *Random Acts of Memory* (1999).

Three Semi-Automatics Just for Fun

The magazine cover girl captured our attention – a sleek, professional blonde in a blue suit holding a small, shiny pistol attractively over her right lapel. Beneath, in bold letters matching her lipstick, were the words: 'Three Semi-Automatics For Fun!'. 'For Fun?' we puzzled, as yet uninitiated to the delights of holster underwear that works; the elegant glamour of a good gun purse and the family fun of mother-daughter hunting – only some of the many topics covered between the pages of this publication. Thus began a guilty obsession with *Women & Guns* magazine,[2] which led, eventually, to a performance piece, *Three Semi-Automatics Just for Fun*. Throughout the process of making the piece we were haunted by the notion of 'fun' both in relation to our own responses to the pro-gun literature and in relation to the construction of our performance. The emphasis on conventional femininity paired with in-depth discussions of high-powered firearms in *Women & Guns* had an inescapably comic effect. Earnest articles appeared each month chock-full of handy, perky tips along the lines of 'while magazines are not hard to load the usual way, women with long nails will appreciate the convenience of using the loading tool'.[3] The juxtaposition, however, of beauty and fashion tips with easy-to-remember jingles such as: 'two to the body, one to the head: guarantees they're really dead' gave the magazines a darkly disturbing quality that we wanted to explore. We knew we would have to work with the black humour already inadvertently present in the pro-gun material in order to confront the issues without trivializing them. Our performance focused first on attitudes, fashions and political trends concerning women in pro-gun publications; and second, on the kind of personal fears and vigilante fantasies that fuel gun-fire. The piece was an exploration of the then current US vogue for pistol-packing wives and mothers as front women of the National Rifle Association[4] and guardians of 'family values'.

Fascinated by the utter contrast in attitudes and laws concerning gun control in Britain and the United States, we decided that in performance we would accentuate our contrasting accents (Helen's English, Leslie's American), and that we would perform the piece in each country, almost as an experiment. The piece premiered in Scotland at the Centre for Contemporary Arts, Glasgow, in 1997 and went on show in the States in the same year. A week prior to the premiere in Glasgow, handguns were unconditionally banned by the British government. While we worked inside the Centre for Contemporary Arts, outside double-decker buses ran along Sauchiehall Street carrying billboard posters advertising a £1000 reward for information leading to the confiscation of newly illegal weapons. Advertising for the performances of *Three Semi-Automatics Just For Fun* had to be removed even from within the CCA building itself due to the colossal number of complaints received from the public – here the danger was in seeming 'pro-gun'. By contrast, when we did the show in Phoenix, Arizona, we were advised to change our publicity image, a 1950s kitchen with three small revolvers resting on the lemon-yellow table, for fear that it was 'too subtle' and 'didn't read'. Several people in Phoenix were shot while we were working on the piece,[5] including a city councillor who was gunned down in chambers for voting the wrong way on a baseball stadium tax, and a young couple who were killed in the middle of the night by bounty hunters with the wrong address. Despite the insane number of local, not to mention national, handgun-related deaths, in Phoenix the danger was in being seen as 'anti-gun'. Audience members in the United States were within their rights to carry concealed weapons into the theatre, raising the stakes to a new high.

The set for *Three Semi-Automatics* was a 1950s kitchen interior replete with a stylish Norge Frigidaire containing an array of pineapple'n'cheese on cocktail sticks, stuck satellite-fashion in grapefruit orbs. In the manner of a sprightly cookery show hostess before a studio audience, the performers, dressed in 1950s costumes and wigs launched into the show:

Leslie: I am myself perfectly willing to take responsibility for the killing I do. I hunt deer, wild pigs and squirrels in the Hill Country around Central Texas. All my children are gun owners and excellent shots. Once I have an animal down, I field dress it myself. There is something very complete about stalking, shooting, dressing and cooking your own food that has been lost in the modern supermarket. I am most proud of my fringed deerskin coat. When people ask me where I bought it, I can tell them I shot it myself.

Helen now performs a buoyant but oh-so-practical DIY demon for the 'studio audience':

Helen: Leftover fruit? Wilted veggies? Overflowing junk drawer? Recycle and have some shooting fun! Nothing is more fun than shooting a new and different target. Here are some inexpensive targets you can make at home. Potatoes explode well, even when hit with a .22 bullet. They can be a bit hard to prop up, but slicing off one end will make them stand firm. Watermelon? It is overrated as a shooting target. You will learn more (and have more fun) from shooting three or four cantaloupes than from lugging one watermelon to the range for a single shotgun blast. A shotgun makes instant slaw out of a head of cabbage tossed into the air. Let your imagination go wild in the produce department of your local store. This is perfect for the broccoli-haters amongst your shooting friends.

In this mode, the ladies wend their way down a delightful path of feminine protection products such as the Diva Gunpurse, 'for a sleek polished look by day and elegant glamour by night' and the Krammer Confidant: 'holster underwear that works'. Tampon-like, the various products promise security, comfort, discretion and the elimination of unpleasant odours, as well as resolving the particular dilemmas of feminine fashion, for example, 'How do you carry a .38 or 9mm handgun without giving up your comfortable elastic-waisted pants, your boxy cropped tops and your pretty dresses?'

The characters/caricatures of the housewives endear themselves to the audience as they extol the virtues of pistols great and small with likeable, earnest vigour. Talk of shot placement and target practice is all tempered by the Doris Day smiles and pearl-necklace normality. The Pillow Pal 'handy holster holder' begins to sound reasonable, even sensible, as do breezy adages such as 'Family Hunting=Family Fun and Family Meals,' or, 'Girls with guns have more fun, and will worry about the hairdos later...' It even starts to sound like, well...fun. Gradually, however, the images of the pistol-packin' homemakers are becoming more and more distorted, the humour darker, the lens more fish-eyed. The audience, previously seduced by the promise of pleasure afforded from the simple comedic juxtaposition of fashion and firearms, rifles and romance, shopping lists and target tips, is now implicated in a far more serious conspiracy with the performers, as the ladies, in a similarly nonchalant fashion, train their sights on the real *raison d'être* of handguns: killing people:

Leslie: Remember, you MUST be prepared to shoot more than once. People who need to be shot, need to be shot soon and often, not later and a little. You will have to keep shooting and shooting and shooting until you are absolutely certain that he is no longer a threat.

Leslie, who has cheerfully illustrated her points by drawing diagrams on a chopping board with a sharp kitchen knife, is now interrupted by Helen who respectfully suggests a previously unmentioned target tip.

Helen: The female gender of adversity adds a whole new dimension to the question of shot placement: pelvic targeting! In short, the technique of firing into a women's lower pelvic region closely approaches the lethality of firing into her chest. In accordance with its purpose of sustaining a foetus for nine months, the uterus is supported by a disproportionately high number of arteries, when compared to other organs. Conversely, any woman sustaining a gunshot wound in that immediate area will lose a large amount of blood, in a very short time.

The above text was drawn from the letters page of *Women & Guns* and had been sent in to the magazine by a female gynaecologist. She had stated that the lower pelvic region was exactly where she'd aim if confronted by a female adversary, explaining that there were no frontal bones that could deflect the lower powered rounds destined for her attacker's uterus. The Editor's eerie response was that her women's pistol league had tried to add this important option to their training programme, but as yet had had no success in their search for female silhouette targets with scoring lines that include the 'Bonus Kill Zone'.

At the end of the first half, light-hearted music strikes up and the ladies drain a couple of quick martinis before tap dancing their way back through the long, narrow, Alice-in-Wonderland-like kitchen and disappearing behind the curtains, leaving the audience alone with their scoring lines. As the audience file out into the foyer for the interval they exchange the playful theatricality of lounge music, blonde wigs and cheese'n'pineapple on sticks for the eerie reality of hard copy. Thirty-six issues of *Women & Guns* magazine, spanning the previous three years, hang in neat eye level rows from the ceiling. Glossy front covers depicting images of perfectly made-up women posing with sleek pistols spin slowly before the faces of the audience, revealing glimpses of the text they have just heard.

Curious echoes of the first act resound through the dark, apparently empty set as the audience retake their places for the second act. Voices are audible, but they are not those of the performers. As they gradually become louder and more distinct, listeners can discern that the voices belong to two older women, one American and one English. In fact, they belong to the performers' mothers who are now repeating the friendly, informative tips and guidelines previously supplied by their daughters. Giant video images of the mothers in their own respective 1990s' kitchens illuminate the large, angular side walls stage right and left, as two spotlights slowly fade up on the performers who are in contemporary dress. There are no tables, no curtains, no cocktail shakers, no heads of cabbage to toss in the air; only two microphone stands, a chair and the apparitions of their mothers floating as outsized spectres on the pale-lemon walls behind them. As the mothers blithely offer advice over the heads of their daughters, 'any woman sustaining a gunshot wound in that immediate area will lose a very large amount of blood, in a very short amount of time', a conspicuously large camera, placed in the front row of the audience, acts with surveillance precision as it tightly frames Leslie's face, which is projected on to the back wall. Framed between the strangely slanted and distorted homemaker's home movies, the giant close up is sharp by contrast, detailed enough to see beads of sweat form on the upper lip. Theatricality, with its luxurious trimmings and comforting boundaries has given way to a severe cinematography. Stripped of their theatrical personae, the performers are left naked in their own images, in turn complicating the role of the viewer/audience. When the performers are acting, the audience is an audience; when the 'acting' stops, the audience find themselves in the position of unwitting voyeurs.

Helen stands with her back to the audience, speaking into a microphone as she questions Leslie, at times referring to a clipboard of questions in her hand, never turning to look at her directly. Leslie sits downstage facing the audience and the camera, uncomfortably boxed in by the close cropping and the tightening circle of questions. In answering each question, she must strain forward slightly towards the awkwardly placed microphone. Initially, the interrogation promises to be fairly innocuous:

Helen: What is your full name exactly as it appears on your passport?
Leslie: Leslie Kay Hill.
Helen: And what does the K stand for Leslie?
Leslie: Kay.
Helen: . . . I see. How tall are you?'
Leslie: 5 foot 6 and a half.
Helen: Exactly?
Helen: What's your blood type?
Leslie: A+.
Helen: Are you eligible to donate blood?'

As a familiar pattern is established, the audience can resume a comfortable performer/audience relationship and the atmosphere relaxes. The interrogator, however, is playing on this false sense of security. The click is almost audible as the safety lock is released:

Helen: Have you ever been followed into an otherwise empty ladies room by a gang of men who were standing in the hallway outside pretending to talk on the payphone? *[long pause]* And did you think it was clever how they came in two by two, leaving one person on guard at all times, pretending to talk on the payphone outside the door?

[pause]

True or false: some people don't deserve to live?
Have you ever been the object of unwanted attention?
Have you ever been singled out?
Followed?
Watched?

The comfortable pattern is disrupted; laughs are half swallowed. Gradually, Leslie is drawn into a series of confessions of personal vigilante fantasies and stories of revenge. The stories place the audience on a precipitous participatory line wherein an enjoyment of the comedy present in the question and answer scenario becomes inseparable from personal accounts of violence and violent fantasy. Leslie, however, slowly redirects the questions and the camera as it stealthily tracks across the stage, focusing in on the new target, zooming in until the back of Helen's head fills the screen. Abruptly, Leslie asks the stage manager to cut the audio/video backdrops of the two mothers who have been demonstrating target practice with Diet Coke cans and heads of cabbage tossed into the air. In an atmosphere of heightened silence and darkness, Leslie instructs Helen to demonstrate for the audience the exact technique she used for looking over her shoulder while walking home at night through the dark, narrow alleyway. Helen's head fills the screen as she turns very gradually to the side and then darts a fleeting, wide-eyed glance backwards. The giant projection of her glance seems to make split-second eye contact with the audience.

Unsatisfied, Leslie asks Helen to step away from the microphone and incorporate walking into her demonstration to provide a more precise reconstruction. A strip of light marks out a narrow pathway alongside the wall and the camera follows her every move as she paces up and down the narrow space and turns her head with a slow inevitability to face what may or may not be behind her. The simultaneous filming and projection creates a double-mirror perspective of infinity, as Helen walks away from the audience towards the screen and her own receding figure. The camera focuses sometimes on a hand as it clenches and unclenches, zooming in on the exposed back of neck, or sometimes letting Helen slip out of the frame before slowly, determinedly, catching up. She turns her head. There is someone there; the camera, the cameraman, the audience. All stalking her. No one is laughing. There is no safe place and after all the 'fun', the loss is felt more acutely. There is an undercurrent of accountability.

Leslie: Have you ever been followed? Watched? Have you ever been the subject of unwanted attention?

The camera too, has turned to look over its shoulder and the audience are now presented, momentarily, with the image of their own faces projected on to the screen in front of them before the room fades to black.

Leslie: And how did that make you feel?

Performing *Three Semi-Automatics* in a State in which 52 per cent of the inhabitants are likely to be armed led to a strange awareness that our audience could potentially be reading the piece 'straight'. Audience notation of the inexpensive and environmentally friendly target tips for future reference would be the ultimate backfire. Likewise, in Glasgow, we faced the danger of being attacked as 'serious' promoters of firearms, at a time when most people, in the wake of the Dunblane school shootings, never wanted to see them again. Portraying the fifties' housewives as accentuated 'feminine' stereotypes with their white gloves, dainty manners and curl-tossin' enthusiasm was a deliberate mechanism set in place to avoid straight readings. The danger of being hoisted with ones own petticoat, however, was real. Interestingly enough, the show did play very much as a 'comedy' in both places. British and American audiences alike engaged with the humour, but who was laughing louder, we wondered, the gun owners or the anti-guns? After the show in Phoenix, we were interested to overhear a snatch of conversation between a man and a woman as they left the theatre, the woman stating with resolve and enthusiasm, 'I'm going out and buying a gun tomorrow.' The man, apparently disturbed, replied sombrely, 'You can have mine.' In the end, we felt that the transient world of the performance offered a useful space in which to question attitudes towards personal fears and violence. The challenge of nonsense inherent in the theatricality of the performance, and the contrasting close-circuit voyeurism, we hoped, allowed for a similar challenge to 'real' publications and perceptions about firearms. While 'pro' and 'anti' gun lobbyists alike may find the black humour irreverent or distasteful, we tend to agree with Jo Anna Isaak in her theory that 'play may well be the most revolutionary strategy available'.[6] And what could be more dangerous, and by implication more *fun*, than playing with guns?

Guerrilla Performance Locator

A web-based Curious piece which explicitly referenced feminism was the *Guerrilla Performance Locator*,[7] an online mapping of global artistic and activist performance and actions. The site took inspiration from the Suffragette movement and on the homepage Leslie Hill delivers a short online movie lecture entitled 'Suffragettes Invented Performance Art'. She relates that many actresses, writers and artists were at the vanguard of the suffrage movement and that they not only understood conventional theatricality but revolutionized 'performance' by bringing it to 'real life'. They coined the phrase 'the personal is political' and created a distinction between theatre and performance – a distinction they made literal through imprisonment and hunger-striking. Suffragettes posted themselves as human letters, performed on flotillas on the Thames during ministers' tea breaks, chained themselves to carriages, synchronized their watches and in unison pulled hammers from their handbags and smashed windows in Oxford Street in a performance precursor to 'I shop, therefore I am'.

The aim of the Guerrilla Performance Locator was thus to encourage viewers to perform acts of resistance – to make spectacles of themselves for things they believed in. Actions and performances were posted on to the locator from across the United Kingdom and globally from as far afield as the Philippines, Malaysia, Santiago and Australia. They highlighted a diverse range of issues from rights for migrant women workers, anti-Iraq war pieces, consumerism, environmental concerns and domestic violence. So while the overall aim of the site wasn't to make a specifically feminist web site – it was to create a cyber-showcase for politically active performance work from around the world – we chose the Suffragette performance activists as our icons, taking inspiration from their willingness to make spectacles of themselves for things they believed in. By the way, you are very welcome to visit or contribute work (feminist or otherwise) to this site: www.placelessness.com/guerrilla

Random Acts of Memory

Often the feminist strands of enquiry within our work – and the work of other female artists – are more embedded, less obvious. For example our performance piece *Random Acts of Memory*, a play on the acronym RAM (Random Access Memory), was a piece which simultaneously juxtaposed and intertwined the live and the mediated formats. So on the most basic level this performance could be seen as primarily an investigation of originals and copies, of live and digital personas. However, on closer inspection the piece was also about notions of reproduction, an issue which has always been associated with feminism. The feminist enquiry in this piece was present not only within the content of the piece but also in the experimentations around form. At the time we made *Random Acts of Memory*, we were working in the largely male domain of interactive technologies at the art-technology lab at the Institute for Studies in the Arts at Arizona State University. Most of the work we had seen using interactive technologies had been developed by male programmers and directors and 'set' on female dancers who never seemed to have a voice, but rather acted as beautiful 'meat puppets' in the work. We were interested in the kinds of live performance work we could make through digital technologies, and how we could play and replay with the direction of the gaze within live performance, using technology to enable a re-viewing and re-focusing of the gaze. Artist and theorist Susan Kozel writes that:

> In the past few years performers and artists have worked to transform our physical and conceptual engagement with technology, well aware of the dangers of simply importing the old gender and racial stereotypes into new media. Is the feminist critic always an outsider? Even in relation to work that transforms the divide between spectator and performer and dismantles the hegemony of the representative process? Much technological performance work thoroughly reworks the trajectory of the gaze in performance using multiple representations of bodies and audience interaction to make subjects *and* objects out of both the performers and audience.[8]

Working with technology allowed us to approach several key questions concerning gender representation, the representation of the female body on stage, the nature of the gaze, reciprocal, voyeuristic. The word 'virtual' dates from the fourteenth-century and comes from the medieval Latin 'virtualis'. The meaning 'so in effect' has developed from 'having power.' Does the 'virtual' performer have more power? Side-stepping the gaze can she redirect it?

In *Random Acts of Memory* the future is digital. The stage environment created reflects a remote and virtual world as the female performers seem almost under house arrest in their state of the art cyber sitting-room. There is no door, cameras and screens are the only portals, there is no way out. Even the large window is revealed to be another screen when the Venetian blinds are opened. A television screen positioned at eye level plays continuously, interlacing 3-D animation of synaptic functions in the brain with 1940s Technicolor footage of Esther Williams, so that synaptic nerves dissolve into scores of smiling synchronized swimmers. The Venetian blinds open and close like mechanized billboards, revealing glimpses, not of a world beyond the window, but of pre-recorded real life and unleashed tension, as the performers ferociously wrestle with each other. This footage is interspersed with live feed of the performance as the performers take turns at recording each other on a digital video camera, set in a prime vantage point usually reserved for the ubiquitous television set, in a 'normal' sitting-room. In a high-tech/low-tech confluence the live feed of the performance jump cuts into pre-recorded digital video, which fades into black and white analogue Super 8 film footage, itself edited digitally. The performers test their memory, claim each other's memories as their own and then forget what they were going to say next.

At the start of the performance the performers sit side by side, discretely copying the movements of the audience members as they take their seats. Dressed alike, in tailored blue suits, each sporting a similar haircut, they are virtually clones of each other. They recount personal memories, which are then recorded and replayed. The memories they earnestly impart are sometimes revealed as not actually being their own, but rather stories they have copied, appropriated from each other, further complicating the relationship between live and mediated. Despite their similarity of their appearance and gesture, the performers rarely interact with each other in the live performance. They do, however, interact with life-size digital clones of themselves which travel through the performance world.[9] The clones say nothing original, speaking only copied words from Hillel Schawtz's *The Culture of the Copy*,[10] that is, copied words about copying. The challenge for the live performers in *Random Acts of Memory* is to try to match the seeming infallibility of their free-floating digital counterparts, their perfect states of being. The performers struggle, forget their lines and grow tense, whereas the clones are word perfect, immaculate, poised. As the performance progresses it seems as if the mediatized world will take over.[11] In a climactic moment of download overload, six Dolly Partons are projected into the space, which then morph, unnervingly almost imperceptibly, into six 'Dolly' the clone sheep. It seems for a moment as if the live performers have been overpowered, engulfed by a pre-recorded world. It is at this point that the live performers exert their own unique capabilities. 'Cut the power', shouts one of the performers, and with a simple toss of a switch, video and film screens flicker and die, Dollies and clones vanish and the live performers, lit by a series of solitary matches, remain. The performers speak fragments of text, stopping abruptly when the flame flickers and dies, recommencing only when a new match is struck. The contrast from the high-tech visuals of before, to this simple, elemental setting that follows is stark and surprisingly satisfying.

The technologies used in presenting *Random Acts of Memory* were the results of research initiatives ongoing in the Technology Development Group of the Institute for Studies in the Arts in Arizona, where *Random Acts of Memory* was devised and premiered. The team had created 'ARVID' (Autonomous Remote Video Imaging Device) a prototype device for human / video performance. Essentially the marriage between Swedish rock'n'roll concert follow spot technology and a purpose built projection system, the ARVID system allowed a spatial interaction between the performer and the video, breaking the boundaries of a fixed projection surface by allowing the video to 'track' the performer in a 360-degree axis. The choice of the technology used in *Random Acts of Memory* reflected a conscious desire on our part to explore some of the anxieties caused by digital technologies as well as purposely choosing to play with some of the boy's toys.

Whose feminist future is it?

So our musings here offer up the idea that you can be a feminist and make work that is allied with feminism in terms of the issues and ideas it engages with, which mingle and mix with other contexts and concerns, sometimes producing overtly feminist work, sometimes producing a piece which is more layered in which the feminist strands don't stand out as clearly. We have said that we imagine the future of feminism in theatre and performance may lie in these hybrid, layered works. There is just one more thing that seems important to point out in relation to our musings – when we're talking about feminism, we are talking about it in a Western context and specifically in a US/UK context, because those are the two countries we come from respectively and the two countries we work in most often. The layered, mixing and mingling of feminism we imagine for the future is only going to be possible if women's rights remain relatively healthy in these countries. If, for example, the Bush administration achieves the coveted right-wing goal of making abortion illegal in the United States once again, you can expect to see some very overt feminist work coming out of the States very soon.

By the same token, the feminist work we have seen during our travels to performance and theatre festivals in India and China, for example, bears little resemblance to contemporary work in the West simply because the reality of opportunity for women and the concerns of feminist agendas in these countries bears little resemblance to the opportunities in the West. In the West, though we forget it all too easily, we are making work in fields that feminists have been ploughing for over 100 years and our hands and feet are a lot less calloused as a result – we can spend a lot less time clearing and a lot more time cultivating, thanks to the women who came before us. An important part of our feminist future will be in remembering and honouring our feminist past and remembering that feminism is an international issue and internationally feminism has a long, long way to go. From now on I'm going to say 'yes' when people ask me if Curious is a feminist company no matter what the show is about.

Notes

1. Definition given by MSN Encarta, http://encarta.msn.com/encnet/features/dictionary, accessed 12 January 2005.
2. Published by Second Amendment Foundation USA.
3. All quotes from shows are from Leslie Hill and Helen Paris unpublished performance scripts.
4. The most powerful pro-gun lobby in the United States.
5. Though not in connection with our performance.
6. Jo Anna Isaak, *Feminism and Contemporary Art: The Revolutionary Power of Women's Laughter* (London and New York: Routledge, 1996), p.3.
7. A 'Shooting Live Artists' commission from BBC1 and the Arts Council of England.
8. Susan Kozel, 'Multi Medea: Feminist Performance Using Multimedia Technology', in Lizbeth Goodman and Jane Du Gay, eds, (London and New York: *The Routledge Reader in Gender and Performance*: Routledge, 1998), pp.299–302, p.301.
9. Isolated through chroma-key shooting, the clones free themselves of the traditional boundaries of the screen not only through isolation, but also through technology of 360-degree axis interactive projection developed by our colleagues at the Institute for Studies in the Arts, Arizona State University, known as ARVID: Autonomous Remote Video Imaging Device.
10. Hillel Schawtz, *The Culture of The Copy: striking likenesses and unreasonable facsimiles* (Cambridge MA: MIT Press, 1996).
11. In Lynn Hershman-Leeson's article, 'The Fantasy Beyond Control', she writes about her interactive video disk called *Lorna* in which the character of Lorna is literally captured in a mediated landscape, see Lynn Hershman-Leeson, ed., *Clicking In: Hot Links to a Digital Culture* (Seattle: Bay Press, 1996), pp.261–73, p.269.

5
'Bad Girls' and 'Sick Boys': New Women Playwrights and the Future of Feminism

Elaine Aston

Feminism has always been concerned with creating a more progressive, democratic society for future generations of women. In the 1970s, a generation of feminist women demanded their liberation from biological and social 'destinies', advocating equality of opportunity, better employment prospects, along with state provision of childcare for those women electing to have children and to work. In the context of a feminist culture and theatre-making, women writers and practitioners contributed to the idea of progressive, 'feminist futures' through, for example, staging resistance to marriage as the only 'career' available to women (*Sweetie Pie*, Bolton Octagon Theatre-in-Education, 1972; *Trafford Tanzi*, Claire Luckham, 1978); educating young women about their bodies and birth control (*My Mother Says I Never Should*, Women's Theatre Group, 1975); and generally stressing the importance of a liberal, progressive education (*Love and Dissent*, Elisabeth Bond, 1983). In brief, whatever its particular route (liberal, radical or socialist), second-wave British feminism cohered around ideas of women's liberation, and feminist theatre engaged with the transformational possibilities that liberation might bring.

However, the idea of a more democratic, progressive future for women was undermined in the 1980s through the backlash against feminism and adoption in the mainstream of a brand of 'feminism' that was bourgeois and individualistic in style. Feminism in the 1980s was in danger of overturning the co-operative and collaborative ethos of the 1970s in favour of a selfish, materialistic creed favouring a minority of already privileged women at the expense of the majority. In theatre, Caryl Churchill's *Top Girls* (1982) famously illustrated the dangers of this new style of 'right-wing' feminism; showed the dangers and consequences for future generations of women of not pursuing a political agenda that took account of both feminism *and* socialism. In *Top Girls* 'superwoman' Marlene's success is at the expense of her slow-learning 'daughter', Angie, while in another of her plays, *Fen* (1983), young girls from an East Anglian Fenland community, who aspire to be hairdressers or teachers when they grow up, will be bound to agricultural (and maternal) labour.

In brief, while second-wave feminism proposed itself as an 'adventure'[1] for future generations of women, it got lost in the 1980s backlash against feminism. 'Somewhere along the line', writes Imelda Whelehan, 'feminism has become the "f-word", perceived to be an empty dogma which brainwashed a whole generation of women into false consciousness of their relationship to power'.[2] Feminism today is represented as outdated and unnecessary: a political movement deemed redundant in a society where women are now understood to have acquired and to be 'enjoying' their equality. Worst still, feminism as it circulates through popular and visual media, is reinvented as a 'free-market' feminism, one characterized as Angela McRobbie explains by 'brutal individualism and the pursuit of wealth and success'.[3] Yet all of this sits uncomfortably alongside the social realities for many women in the twenty-first century, for whom inequalities persist and whose lives have been damaged, as Churchill among others prophesied, by the disappearance of a socialist and feminist agenda.

Indeed, in contradistinction to the idea of a postfeminist society, social realities would seem to argue that the idea of a more progressive future, for women especially and for society generally, is further away from our grasp than it was in the 1970s. Feminist playwrights from that era writing today, such as Caryl Churchill, Timberlake Wertenbaker or Bryony Lavery, present us with drama that grows increasingly dark: where welfare systems are ailing and children are at risk (Wertenbaker, *The Break of Day*, 1995); young girls are prey to child killers (Lavery, *Frozen*, 1998), or young women grow up conditioned into actively participating in a world of violence and terror (Churchill, *Far Away*, 2000). Writing by this generation of second-wave feminists is fearful of a future that is not being shaped by the progressive and democratic vision that feminism previously had to offer.[4] If this is the outlook that obtains for a feminist generation of playwrights, then what about women playwrights who are just beginning careers in theatre? Do these writers fictionalize worlds that connect in any way to a feminist tradition? Are they writing optimistically about women's lives, or even dramatizing women's lives at all? This is the focus of my essay:

exploring what a younger generation of playwrights has to say, and whether in turn this can be instructive or insightful for a revitalization of a feminism that remains invested in a set of progressive, democratic principles, as opposed to its reactionary free-market 'sister'.

To pursue this discussion I draw on a snapshot of contemporary women's playwriting from the 2003 autumn seasons at the Royal Court and Bush Theatres, two of London's premiere venues, both of which are noted for their support of new writing which might be roughly characterized as counter-cultural. In the past, notably under Max Stafford-Clark's directorship (1979–93), the Court nurtured the careers of feminist playwrights such as Churchill, Daniels and Wertenbaker, while the Bush developed a long-standing relationship with Catherine Johnson.[5] Turning to the present, analysis in this essay draws on professional debut plays by three women dramatists: Stella Feehily *Duck* (Out of Joint touring production and Royal Court, November 2003); Emma Frost *Airsick* (Bush Theatre, October 2003), and Lucy Prebble *The Sugar Syndrome* (Royal Court, October 2003). Each of these three plays examines issues that may be helpful to redirecting or reinvigorating feminism.

'Bad Girl' drama: *Duck*

Core to the principles of second-wave feminism was its objection to the objectification of women. Protests against the Miss World beauty pageants held throughout the 1970s and other similar campaigns fuelled feminist argument about ways in which women were coerced into or constructed by and through ideologies of femininity. More recently, 'new feminism' of the 1990s has accused second-wave feminism of putting too much emphasis on sexual politics and culture: of concentrating predominantly, if not exclusively, on body culture at the expense of material concerns.[6] More worrying still, however, has been the reclamation of the feminine through the 1990s explosion of girl culture and power: the idea that girls can readily access power through a sexualized feminine. Dubbed by some as the 'feminism' of the 1990s,[7] 'girl power' is more accurately to be understood as an individualistic style of self-promotion: one which encourages girls to believe that self-confidence and sexually aggressive behaviour is a means to empowerment: a means of getting on and getting what you want. Girls adopting bad-boy behaviour, being tough, glamorous and sexually aggressive, featured heavily in the British media during the 1990s.[8]

However, this popularized, reactionary form of 'girl power' is as dangerous as its 'top girl' predecessor. Both carry the promise of empowerment whilst concealing the realities and disadvantages of social and cultural systems that militate against the fulfilment of this promise. Taking issue with this anti-feminist version of 'girl power' is a tradition of hard-hitting, brutal drama by women that adopts a very different attitude to girl culture. In the 1990s, playwrights Rebecca Prichard and Judy Upton were notable 'champions' of disadvantaged communities of young girls: girls dropping out of school and having babies (Prichard, *Essex Girls*, 1994); or girls rejecting or leaving home and taking to a life on the streets in girl gangs (Prichard, *Yard Gal* 1998; Upton, *Ashes and Sand*, 1994).

The trend for this hard-hitting, 'bad girl' drama is one that I would argue is continuing, as evidenced in the first of the three plays under consideration here: the debut drama *Duck* by Irish dramatist Stella Feehily.[9] Directed and toured by Max Stafford-Clark and his company Out of Joint, *Duck* further reflects Stafford-Clark's enduring support for work that gives dramatic representation to young women of working-class, or more recently under-class, communities.[10] While *Duck*, like other of the 'bad girl' dramas, does not purport to an explicit engagement with feminism, it nevertheless connects to feminist interests. Specifically, I would argue this connection to feminism on account of its dramatization of a resistant feminine that takes issue with the dominant view of 'girl power'; for its dramatic focus on girls' friendship; and last, but not least, for its implicit criticism of an academic feminism that is perceived to be out of touch with the lives it claims to be 'theoretically' interested in.

Set in Dublin, *Duck* focuses on the exploits of two teenage girls, Sophie and Cat (known as Duck). These are girls who have been dealt a rough deal and who want more out of life than life seems to offer them. The teenage girls in *Duck* struggle to make something of their lives despite their respective dysfunctional families and, in Cat's case, a sexist and physically violent relationship with her boyfriend. Femininity in their working-class culture is not rejected (as it was by a feminist generation), but is messed up or roughed up through aggressive masculine behaviour: swearing, fighting and street crime. In the opening scene, for example, Cat and Sophie are out on the street late at night. Cat has torched her boyfriend's jeep on account of how she was tired of being left in the car while he got on with his 'business' (drug dealing). The girls have all the trappings of femininity: short skirts and high-heeled shoes, but they are also drunk, foul-mouthed and well practised in the art of street fighting. In this opening scene they see off two would-be male attackers, verbally and physically (cutting the boys with a broken bottle), the kind of hooligan behaviour traditionally associated with boys rather than girls. While there are a few oblique references in

Duck to the ongoing 'Troubles' in Northern Ireland, the emphasis in the drama is on mapping cultures of violence with gender warfare. Femininity in this street violence functions as a Trojan horse: feminine appearance conceals a surprise attack on ill-prepared, unsuspecting male victims.[11]

Much of the contemporary 'bad girl' drama is, as *Duck* exemplifies, characterized by a disturbing note of female aggression: young women challenging male sexual dominance, verbally, sexually and physically. Not that female aggression serves to better the girls' lives or fulfil their dreams – quite the opposite in fact, as more often than not violence gets them deeper into trouble (Cat always comes off the worst in fights with her boyfriend, for example). The representation of violence among young women does, however, signal a culture of resistance born of the increasingly violent 'rub' between aspirations and social realities. Violence erupts precisely because these girls are not (as the girl-power myth suggests) getting what they want. Unlike a 1970s generation of women fighting for equality and rejecting femininity as a site of oppression, recent outbreaks of girls behaving badly,[12] suggest a generation of young women who assume their right to equality and wield (rather than reject) the feminine as a social and cultural weapon in wars that are both gendered *and* classed.

Core to the dramatic composition of *Duck* is the friendship between the girls Sophie and Cat. In second-wave feminist theatre women's friendship was foundational to building 'alternative' female communities, potentially resistant to dominant and oppressive heteropatriarchal systems. Mirroring the experience of feminist consciousness-raising groups, dramatic representations of an all-female community created the 'space' for women characters to air grievances about male partners, fathers or other male authority figures, held responsible for the various discontents and problems in their lives, as in Pam Gems's *Dusa Fish, Stas & Vi* (1976), or Nell Dunn's *Steaming* (1981). Female friendships and communities within this kind of explicitly feminist drama argued the personal as political: personal problems were not represented on an individualistic basis, but were demonstrated as causally linked to the patriarchal and capitalist organization of society.[13]

By contrast, the more recent 'bad girl' drama by women playwrights has represented women's friendship as foundational to the formation of an oppositional or resistant community that is not rooted in the idea of 'women coming together', but of women 'ganging' up to fight their corner. Prichard and Upton's girl gang plays of the 1990s, for example, promote an idea of bonding through girls behaving badly together.

In *Duck* the friendship between Cat and Sophie includes bad-behaviour bonding (as evidenced in the play's opening scene as previously described). The friendship also functions, however, as a means of working out individual and shared difficulties: of dysfunctional families (both girls have them) or sexist relationships (as in Cat's case with her nightclub-owner, drug-dealing boyfriend). While traces of a legacy of an earlier set of feminist concerns might be argued for in this friendship device, more significant is what can be understood from the way in which these young women are positioned as *strangers to feminism*.

In the absence of feminism or any kind of cross-generational support (if anything the older generation tend to be obstructive rather than supportive),[14] the girls resort to educating themselves. Sophie, studying at University College Dublin, and learning about history, connects her historical studies to their contemporary lives. Specifically, she draws Cat's attention to the parallel between two rebellious women figures in seventeenth-century history, one a scold and the other an arsonist, and their own experiences. What Sophie and Cat have in common with this seventeenth-century female ancestry is their verbal protesting (the way they talk back, answer back) and in Cat's case, arson: the torching of her boyfriend's jeep.

Their favourite, arguably their only, space for 'learning' in *Duck* is the bathroom, an intimate, private space, but also a site of sexual violence: Cat, for example, receives a violent and repeated 'ducking' by her boyfriend in the bath (*Duck*, scene 12). While the girls' history 'lesson' takes place in the public toilets of a nightclub, in another scene Sophie locks herself in the bathroom at home to dictate an answerphone message for Cat. The messages consists of Sophie reading aloud and listing dictionary definitions associated with the female sex, looking up gendered words that have become terms of sexual abuse – ones that the girls are very familiar with. As Sophie quotes archaic definitions of words, it serves as a reminder that sexual abuse links to centuries of women being badly treated.

Cat is antagonistic towards Sophie's history lessons, however, and her friend's idea that she might want to 'think about' these issues:

Cat: What's to think about?
Nothing's forever.
No job,
No life.
I could easily get blown up.

(*Duck*, scene 9, p.59)

In marked contrast to feminism's own investment in knowledge of oppression, Cat refutes the idea that 'thinking' about her condition is in anyway valuable or helpful. Moreover, her ensuing question to Sophie, 'Will you ever stop trying to explain me away?' (*Duck*, scene 9, p.58), articulates a sense of danger or anxiety for the 'subject' taken up in analysis or theory.

In brief, this pivotal scene from the play touches on issues that circulate in various strands of feminist, queer, or postcolonial theory, where the anti-essentialist thrust of the theoretical is subjected to interrogation or reconsideration (see discussion of this point in Chapter 11). As McRobbie argues, 'while there has been an enormous output of feminist post-structuralist writing of late, there has been some resistance to looking outside theory and asking some practical questions about the world we live in'.[15] 'Practical questions' are what Cat insists upon rather than textbook explanations that cannot explain away, do away with, or connect to the extremities and dangers of her lived experience.

Implicated in all of this is feminism's failure to connect to the lives of present and future generations of women. For a feminist generation of women struggling for 'affirmation of themselves' there was, as socialist-feminist practitioner Mary McCusker of Monstrous Regiment theatre company described, the recourse to feminism as the solution to your problems: 'If you were fighting your way out of something in the 70s there was feminism there for you to find.'[16] As a feminist spectator, I found myself concerned about the absence of a political landscape that might assist the girls in 'fighting [their] way out'. To elucidate further: in the play's final scene, I felt the tension between the girls' desire to escape and the ambiguity that hangs over this as a realizable outcome. In the final scene, the girls are out on the street with several of Sophie's possessions (she has left home), waiting for a taxi that has not showed up. 'The Great Escape this ain't', says Cat (*Duck*, scene 20, p.109). Leaving the Court studio at the end of the performance, a woman behind me turned to her friend to ask her what the girls were doing at the end? Where were they going? The ambiguity of the final image of the two girls waiting for a taxi they are not sure will arrive is striking. It balances hopeful expectation with a feeling of the girls not so much having somewhere 'new' and different to go, somewhere where their lives will be better, but with the sense that they have been 'here' before. Arguably, I 'read' into this my own position as a feminist spectator: the sense that I have been here before and the mixture of disillusionment that so little has changed, but also the prospect, anticipation and expectation the comes from *feeling* the moment in which lives might be about to be lived differently.

'The decline of the masculine empire'?: *Airsick*

One of the factors militating against the girls' 'escape' in *Duck* is the persistence of gender inequality allied to their particular social and cultural circumstances. Men are represented as relatively privileged in both economic and sexual terms, even if that privilege is attained through criminality (Mark's drug-dealing business) and/or violence (Cat's relationship with Mark is violent and abusive, while her affair with Jack, that appears to be more caring and supportive, proves to be nothing more than an older man's infatuation with a younger woman's body). In brief, this points to the issue of male power as an enduring feminist concern.

Second-wave feminism identified men, privileged by capitalist and patriarchal systems, as largely responsible for women's oppression. For many feminists, radical feminists especially, men were *the* problem women needed to address. Looking back on and assessing the advances of second-wave feminism in the early 1980s, activists Anna Coote and Beatrix Campbell noted that in the interests of feminism's future, attention *still* needed to be paid to the issue of 'male privilege', endorsed by and promoted through systems of education, media and mass culture.[17] Again, despite more recent claims by Naomi Wolf and others to the 'genderquake' of the 1990s and 'the decline of the masculine empire',[18] the signs are that the 'empire' is by no means 'lost', but is in trouble, ailing and 'sick' in a way that is infecting and affecting women's lives. Hence, various 'new' feminisms of the 1990s, each with its own critical take on 1970s feminism and argument about where feminism should be going next, nevertheless arrive at some sort of consensus when it comes to the issue of male privilege, whether this is argued in terms of the social and cultural fabric of Britain, media and popular culture, or technology.[19]

While feminist theorists and critics argue an enduring need to attend to male privilege, men, on the other hand, have been busy blaming feminism for a perceived loss of privilege and status. Throughout the 1990s, 'boys in trouble' was a dominant thematic in British cinema (in films such as 1997 hits *The Full Monty*, or *Brassed Off*), and in television where a 'boyish gang mentality' featured in a variety of programmes such as *Men Behaving Badly, Fantasy Football League* or *They Think It's All Over.*[20] In theatre in the 1990s, playwright David Edgar argued 'masculinities and its discontents' as an 'overarching theme' for a number of male playwrights, writing from a diverse range of sexual, cultural or political perspectives.[21] It does not automatically follow, however, that attention to masculinities is necessarily progressive in outlook. As Claire

Monk persuasively argues in the case of British cinema in the 1990s, a 'post-feminist male panic' influenced a 'strand of male-focused films whose gender politics were more masculinist than feminist'.[22] Typically, a backlash response to feminism has been to reposition men in the victim position that feminism previously argued for women.[23] Scapegoating feminism for the boys' troubles, however, tends to elide the complexities of social and cultural change, especially with regard to economy and employment. 'The betrayal of modern man', as feminist Susan Faludi identifies and argues, is consequent upon the 'larger culture'. The danger is, Faludi explains, that 'by casting feminism as the villain that must be defeated to validate the central conceit of modern manhood, men avoid confronting powerful cultural and social expectations that have a lot more to do with their unhappiness than the latest sexual harassment ruling'.[24] What is called for is an understanding of the 'larger' cultural and social picture that conditions masculinities, male anxieties and behaviour, and shapes men's sexual and familial relationships with women. A critical, wide-ranging look at modern masculinities and their attendant, damaging consequences for women is dramatized in Emma Frost's *Airsick*.[25]

Airsick opens with a framing device that twins the birth of Lucy, daughter of Mick, to American scientist John Wheeler's naming of Black Holes in 1968. In the play, Black Holes figuratively come to stand for creative and destructive forces that are represented as gendered. The young generation of men in the play – Lucy's American boyfriend Joe and Gabriel, a young man she meets while travelling – figure as destructive forces orbiting around Lucy. Joe, a risk management consultant, proves selfish and uncaring towards her. Gabriel, with whom Lucy eventually has a one-night stand, following the break up of her relationship with Joe, actually brings about her death: fatally infects her with hepatitis B.

Rather than looking to feminism as the root and cause of all men's problem's, *Airsick* dramatizes a more complex view of masculinities: one that gives consideration to changing employment patterns and a generalized anxiety about how people are making their futures through work and relationships. With regard to work, Lucy, as an artist, is the only person involved in making something: Joe manages risk (insurance); Gabriel and Scarlet, Lucy's best friend, are both itinerant workers in the casual labour market. Where the figure of the artist in the past has been romanticized as a penniless creator of works of art, New Labour's fashioning of the creative industries, that began in the 1990s, has promoted culture and those who make it as central to the future of the British economy.[26] Realigning industry with creativity displaces a nineteenth-century idea and legacy of industrial production and its emphasis on *man*ual labour and *man*ufacturing. As Claire Monk argues 'the structural changes of the post-industrial era had, by the 1990s, virtually obliterated two employment strata overwhelmingly dominated by men – unskilled workers and middle-management – from the workforce' ('Men in the 90s', p.159). As a *woman* artist, Lucy figures the feminization of future economic prosperity that fuels male anxiety. This feminization is further heightened as, unlike corporate business, creative industries rely on small-scale, individual artistic initiatives that arguably have more in common with a pre-industrial cottage industry: working out of home 'premises' with the promise of future careers and financial success. Lucy's art commissions, for example, bring prestige rather than wealth, though with an expectation that recognition will eventually bring financial reward.

The male characters in *Airsick* are envious of Lucy's 'career' as a 'maker',[27] given their own feelings of dissatisfaction (Joe) or non-belonging (Gabriel). Dissatisfaction with employment is also bound up in personal and sexual relations. Joe, through his relationship with Lucy, relocates from New York to London because his career in risk management is less successful than those of his American peer group. 'I was even pleased about nine eleven', Joe confesses to Lucy during the break-up of their relationship, 'Because at least it took some of them out of the running. At least it opened up a chance for me' (*Airsick*, Scene 10, p.64). This shocking admission links global terror to the emotional, personal terror of heterosexual relations: men fear they will be destroyed by women,[28] and women in turn get 'sick', are infected by 'diseased' masculinities.

While men appear incapable of responsible behaviours towards women, women in turn have internalized a negative image of themselves. In a series of monologues that punctuate the realistic scenes of the play, Scarlet, abused by her father and countless others, narrates a damaged sexuality and repressed anger. Lucy also regards her father as in part responsible for her lack of self-worth and poor sense of body image (which Joe also encourages). The self-portrait, for example, that Lucy talks to Gabriel about, one that she had exhibited in New York and one that occasioned her meeting Joe, is made out of rubbish.[29]

In questioning paternal behaviour, *Airsick* brings a further, cross-generational dimension to the play's take on masculinities. Mick, the father, belongs to a generation of men promised adventures they did not get to have and the play's space motif is used to give expression to this. Identified with science and space, Mick is a man with a failed marriage behind him, and an ineffective, if not damaging, father. As Judy Wajcman observes, space travel was the dream of the 'modernist hero': a particular generation of men set to conquer other 'worlds'.[30] The patriarchal figure 'trapped' into a personal, parental and familial matrix is the failed 'modernist hero' – one who coveted but did not get to have adventures of scientific discovery and whose disappointments damaged marital and filial relations.

The space motif also figures the play's overarching idea of destructive relations between the sexes. Destructive objects, so scientific theory proposes, are born from the deaths of the brightest stars, eventually pulled into the gravity of the Black Holes, whose danger is unavoidable because it is undetectable: you do not seeing it coming. Lucy, the bright star, is destroyed by the men around her: by a father who fails to care for her as a child, which in turn plays into her choice of dangerous and ultimately fatal relationships with male partners. When you choose a relationship, you are also choosing a 'future', but one that can throw you completely off course. Previously, a feminist politics offered a guiding principle to women that enabled them to choose lives and relationships differently: with a greater desire for equality – though as the suicide of the lovelorn, feminist activist Fish in Gems's *Dusa, Stas, Fish & Vi* (1976) cautioned, this did not always work. Yet, if politics cannot dictate the personal, it may assist with keeping relations on a more progressive course, and in the present moment, *Airsick's* dramatization of the 'deadly embrace' of heterosexual relations (*Airsick*, scene 6, p.23), makes a persuasive argument for this as an urgent site of feminist re-evaluation and intervention.

In brief, Gabriel, the international traveller, posits the idea of contemporary heterosexuality as a lethal, fast-spreading virus that is not quarantined and puts women at risk. He personifies the deadly and fatal irresponsibility of successive generations of men towards women: conquering 'heroes' harmful to the futures of both sexes. Equally, *Airsick* illustrates the need for young women to address the negative view they have of themselves, one that they have been socially and culturally conditioned into believing, in order, as Scarlet suggests, to be able to go back to 'a place where other futures [are] still possible' (*Airsick*, scene 13, p.70).

The sexual 'freedom' of cyberspace: *The Sugar Syndrome*

While *Airsick* makes reference to a contemporary transatlantic landscape, increasingly at risk of terrorist attack and anxious about its future, it specifically figures risk, terror and anxiety as a product of damaged heterosexual relationships destroying both private and public spheres: personal, sexual relationships, as well as the production of (creative) labour. In both *Duck* and *Airsick*, it is the girls who are placed at greater risk from unreconstructed and/or disaffected masculinities that desire to own and objectify the feminine, or destroy it rather than risk contamination. While the desire for 'flight' is common to both plays, gender forces play a significant role in keeping the girls grounded in *Duck* or 'crashing' in *Airsick*.[31] In the third and final play for discussion in this essay, Lucy Prebble's *The Sugar Syndrome*, an idea of escaping damaged subjectivities and sexualities is linked to new technologies and cyberspace: to the possibilities of electronic lives and on-line identities that may help to re-vision the pain and damage of socialized, gendered bodies.

In a drama reminiscent of Churchill's debut writing (specifically *Lovesick's* dramatization of people falling love with or desiring the 'wrong' person and *The Hospital at the Time of the Revolution's* motif of a daughter force-fed values she cannot stomach),[32] Prebble's *The Sugar Syndrome* dramatizes a series of electronic encounters in an unlikely 'love' triangle. In cyberspace, Dani meets Lewis and Tim. Lewis is a young man (probably a virgin) who asks her to meet him for sex. Tim is an older man, a sex offender who desires little boys. In an Internet chat room Tim believes Dani to be an 11-year-old boy, not a 17-year-old girl. Nevertheless, Tim and Dani strike up a friendship out of their respective 'abnormalities': Dani is recovering from an eating disorder and has only recently rejoined her family after a spell in a clinic; Tim has been in prison and is trying to make a life for himself, which is difficult given that he is on a sex-offenders register. Both Tim and Dani face a daily struggle of not giving in to their respective disorders: Tim's passion for young boys, Dani's eating problems.[33]

Unlike Tim and Dani, Lewis, as a young man interested in sex, appears to have a 'normal' appetite. Positioning Lewis in the triangular relationship with Dani and Tim, however, questions the normalization of his sex/gender behaviour. In particular, it illustrates ways in which women may be complicit in assuring men of their sexual pleasure to the detriment of their own. For example, when Dani first meets Lewis for sex, Lewis has to apologize for the size of his cock, about which he had apparently exaggerated in cyberspace. (The Court audience laughed at this confession: the laughter of recognition at a lie so frequently told.) Dani 'talks' sex to Lewis, makes him feel like the 'big' man he wants to be, replicates the fiction of the chat room, but in a way that negates her own sexual pleasure: '*Dani's eyes roll heavenward as he [Lewis] touches her, more through annoyance than pleasure*' (*Sugar Syndrome*, Act One, Scene 1, p.6). In performing or acting out the role required of her as desired object, Dani is alienated from the site of her own desire, while Lewis comes to believe (obsessively) in the possibility of his.

My question that follows from these observations is whether the playing out of sexualities on- and off-line in *The Sugar Syndrome* is in any way helpful to a feminist agency? Does it, for example, suggest a means of agency or escape from the gender wastelands encountered in *Duck* and *Airsick*?

While science and technology have the capacity to change lives and futures, feminism has long since recognized that new knowledge is determined and framed by the values of those who advance and invest in it. Advances in biotechnology, for instance, where women have much to gain, are also viewed with resistance, scepticism and apprehension by feminism, given the male domination of biological science.[34] Excited by the idea that the virtual bodies of cyberspace might emancipate bodies from the damaging effects of gender difference and heteronormativity, feminism has, as Wajcman explains, given new technologies a relatively enthusiastic reception. Nevertheless, Wajcman cautions that a 'focus on cyberspace as the site of innovative subjectivities that challenge existing categories of gender identity may exaggerate its significance' (*Techno Feminism*, p.75). What *The Sugar Syndrome* dramatizes is an idea of cyberspace not so much as a space for breaking down gender categories as a reflective space which helps to heal the pain and damage of the socialized, gendered body. For Dani, making intimate contact in cyberspace serves as an antidote to her inability to be intimate with her family and substitutes for her lack of close friendships. Life on-line is ultimately not an escape into a utopian elsewhere, but a conduit for dealing with the here and now of gender damage.

The Sugar Syndrome also cautions that cyberspace may be a source of gender damage to which critical attention should be paid. Feminism is accustomed to intervening in familiar cultural forms and their gender relations and connections to women's lives: has studied women's magazines and their representations of an 'ideal' feminine (a constant source of anxiety for Dani given her battle with anorexia), and has a history of objecting to the objectification of women in all kinds of contexts from tabloid pictures (in the kind of newspaper that Dani's absentee father owns) to the pornography industry. The play suggests that cyberspace, as a new technologized system of culture and communication, requires a similar kind of feminist attention and vigilance. Electronic intimacy has its sexual pleasures (the web is, as Dani argues, 'a place where people are free to say anything they like. And most of what they say is about sex' (*Sugar Syndrome*, Act 1, scene 2, p.8)), but it is also open to abuse: the Internet gives Tim access to child pornography and it allows Lewis to harass Dani on-line.

The play's closure strikes a hopeful, albeit fragile, note: Dani returns to her mother, Jan, not in the kind of mother–daughter reconciliation characteristic of second-wave feminist drama, but as a moment in which two women, of different generations, address the damage in their lives that prevents them from moving on: both the hurt they have caused each other, and the harm inflicted through their respective sexual relationships (Dani with Lewis; Jan with her husband). Symbolically, Dani also burns her 'Thinspiration' book of magazine images and destroys Tim's laptop which stores child pornography. In brief, moving on requires an admission of familial, marital and sexual damage, and the removal of cultural values and means of communication that coerce and condition oppressive futures. None of which, however, is easy. Visually, in the design of the stage space where minimalist domestic interiors multi-locate and transform into the 'outside' world of cyberspace, the cross-infection of interiors and exteriors images the 'baggage' of the past that remains a risk to the future.

Conclusion: broken realism and connections to feminism

Common to all three plays is their style of broken realism: a realism that reflects a being-in-the-world-that-is, fractured by a desire for other lives, other worlds. Encountering this kind of broken realism in the plays, rather than more experimental forms of new writing,[35] in some way suggests a break with a feminist theatre tradition that locates in dramatic forms that are capable of containing or presenting a vision of a radical, more progressive future. From most feminist perspectives, realism has been viewed as 'deadly'.[36] That said, I would argue that the current style of fractured realism among new women dramatists, formally reflects the tension and frustration arising out of a contemporary moment in which a generalized myth of postfeminism collides and is in conflict with particular social, sexual and cultural experiences.

In production terms, this tension is most keenly felt in moments when realism's registers of performance are heightened to the point that dramatic fictions threaten to fracture into dangerous realities. A spectator *feels* the violence of the intimate bathroom scenes in *Duck*, for example, because realism's conventions of pretending (acting) are at risk; the rules of mimetic playing may be broken as a *real* body, ducked and repeatedly ducked, is at risk. Or, acting out Dani's binge eating and vomiting in *The Sugar Syndrome* compels the spectator to *see* the anorexic body in pain. Antithetical to a Brechtian presentational mode of seeing, that feminist theatre previously adopted to serve its political interests, broken realism politicizes its spectators through playing by its own mimetic rules: showing the world that is, to the point where it cracks, crashes and is torn apart by the sheer weight of its own violence.[37]

Previously, it was feminism's task to make visible the violence done to women: to expose the 'naturalized' damage done to women's lives. This snapshot of new writing suggests that it is now the damage done to feminism that is being exposed: revealing the contemporary myths of girl power, the backlash crisis in masculinity, and, overall, the consequent alienation of feminism from the future generations of women with whom it had hoped to connect. The energy of feminism's second wave came from women's growing awareness and intolerance of centuries of oppression. A new or renewed wave of energy may come, this new women's drama suggests, from the connections young women make to feminism as the 'place' to go back to that makes 'other futures still possible'.

Notes

1 See Angela McRobbie and Trisha McCabe, eds, *Feminism for Girls: An Adventure Story* (London: Routledge, 1981).
2 Imelda Whelehan, *Overloaded: Popular Culture and the Future of Feminism* (London: The Women's Press, 2000), p.16.
3 Angela McRobbie, *Feminism and Youth Culture*, 2nd edn (Basingstoke: Macmillan – now Palgrave Macmillan, 2000), p.211.
4 I pursue this point elsewhere in *Feminist Views on the English Stage: Women Playwrights, 1990–2000* (Cambridge: Cambridge University Press, 2003).
5 For a list of other women playwrights at the Bush, see relevant entries in Mike Bradwell, ed., *The Bush Theatre Book* (London: Methuen, 1997), 'Bush Highlights – An Annotated History', pp.8–42. For plays by women writers Lesley Bruce, Catherine Johnson, Tamsin Oglesby and Naomi Wallace, see *Bush Theatre Plays* (London: Faber & Faber, 1996).
6 See, for example, Natasha Walter, *The New Feminism* (London: Virago, 1998), ch. 1, 'What is the New Feminism?', pp.1–9, p.4. Walter claims, wrongly in my view, that 1970s feminism neglected 'material inequalities' in favour of the cultural (p.3), though this fails to recognize the strong connections that second-wave British feminism had to socialism.
7 See comments from the all-girl band, Spice Girls, quoted in Whelehan, *Overloaded*, p.45.

8 For media examples and discussion of bad girl behaviour, see Whelehan, *Overloaded*, ch. 2: 'Girl Power?', pp.37–57.
9 Stella Feehily, *Duck* (London: Nick Hern Books, 2003).
10 In 1980, for example, Stafford-Clark brought Andrea Dunbar's debut play, *The Arbor* to the Court: a drama that takes a hard look at the brutal deprivation of life on a Bradford council estate as experienced by a young woman, whom Dunbar presents quite simply as 'The Girl'. Dunbar died at the age of twenty-nine having written three plays. Stafford-Clark recently revived her second play *Rita, Sue and Bob Too* (1982; 2000), which Out of Joint toured with *A State Affair* (2000) – a new play, devised out of a trip back to the council estate that originally inspired Dunbar's *The Arbor*.
11 See also the opening scene of Upton's *Ashes and Sand* in which young girls gang up and mug an unsuspecting male victim, beguiled and disarmed by their sexual advances.
12 See my discussion of girl gang culture in *Feminist Views*, ch. 4, 'Girl Power, the New Feminism?', pp.59–76.
13 Some feminist playwrights productively complicated the 'universalizing' impulse of this friendship through the inclusion of differences. Sarah Daniels staged the difficulties of realizing friendships and relationships across differences of class, sexuality and generation, while Caryl Churchill famously located class conflict in the 'sister' relations between working-class Joyce and 'top girl' Marlene in *Top Girls*.
14 During the course of the drama, Cat has an affair with an older guy, a man old enough to be her father (a point reinforced by doubling the roles of father and lover), who is a successful writer. Educated, cultured and not the violent type, Jack, the writer, appears, briefly, to be a possible educator and a better prospect than her boyfriend (though ultimately, she is proved wrong about this). The classical music and learned books Jack has in his home are, as Cat observes, different to her mum's 'library' – a 'mixture of Catherine Cookson and Harold Robbins' (*Duck*, Scene 6, p.41). The problem, however, is that Jack's interest in Cat is purely sexual, whereas Cat desires a relationship that brings romance, love and financial security.
15 Angela McRobbie, *In the Culture Society: Art, Fashion and Popular Music* (London: Routledge, 1999), p.75.
16 Mary McCusker, quoted in Elaine Aston, ed., *Feminist Theatre Voices* (Loughborough: Loughborough Theatre Texts, 1997), p.65.
17 See Anna Coote and Beatrix Campbell, *Sweet Freedom: The Struggle for Women's Liberation* (London: Picador, 1982), ch. 9, 'The Future', pp.235–48.
18 Naomi Wolf, *Fire with Fire: The New Female Power and How it Will Change the Twenty-First Century* (London: Vintage, 1994). I am quoting from Wolf's title for Part I of her study: 'The Decline of the Masculine Empire: Anita Hill and the Genderquake'.
19 For discussion of each of these three areas respectively, see Natasha Walter, *The New Feminism*; Imelda Whelehan, *Overloaded*; and Judy Wajcman, *Techno Feminism* (Cambridge: Polity Press, 2004).
20 See Whelehan, *Overloaded*, 'Lads: The Men Who Should Know Better', p.64.
21 David Edgar, ed., *State of Play* (London: Faber, 1999), p.27.
22 Claire Monk, 'Men in the 90s', in Robert Murphy, ed., *British Cinema of the 90s* (London: *BFI*, 2000), pp.156–66, p.157.

23 For backlash publications that rehearse the men as victims of feminism argument, see Neil Lyndon, *No More Sex War: The Failures of Feminism* (London: Sinclair-Stevenson, 1992) and David Thomas, *Not Guilty: In Defence of Modern Man* (London: Weidenfeld & Nicolson, 1993). For a discussion of men on the defensive in a theatre context, see Michael Mangan, *Staging Masculinities* (Basingstoke: Palgrave Macmillan, 2003), ch. 7, 'Contemporary Masculinities', pp.207–48.
24 Susan Faludi, *Stiffed: The Betrayal of Modern Man* (London: Vintage, 2000), p.14.
25 Emma Frost, *Airsick* (London: Nick Hern Books, 2003).
26 See McRobbie's discussion of culture as 'the engine of economic growth' (*Culture Society*, p.3).
27 Lucy calls herself a 'maker', explaining that she does not 'like the word "artist". It sounds... too important' (*Airsick*, Scene 2, p.8).
28 Mick (Lucy's father), for instance, is obsessed with the idea that women will infect him with a sexual disease.
29 See *Airsick*, Scene 4, p.18. Lucy making art out of discarded objects also reflects on the 'waste' produced by a contemporary consumerist society.
30 Wajcman, *Techno Feminism*, p.56.
31 Lucy's death is prefigured in the play's opening scene which re-enacts a childhood party game; Lucy, blindfolded, is surrounded by three men, one of whom is her father, who play a game of make-believe flight, terrifying the 'child' into believing she is on an aeroplane that is breaking up and about to crash.
32 Both of these can be found in *Churchill: Shorts* (London: Nick Hern Books, 1990).
33 Being sensible, not giving in to urges and going for quick 'fixes' relates to the title of the play. *The Sugar Syndrome* refers to post-war times when people preferred to use their rations for a quick sugar treat rather than saving them up, sensibly, for meat products (see Jan's speech, *Sugar Syndrome*, Act 2, scene 5, p.71).
34 This is an enduring concern of much feminist drama. See, for example, Sue Townsend's dystopic *Ten Tiny Fingers, Nine Tiny Toes* (London: Methuen Drama 1990), or Timberlake Wertenbaker's *The Break of Day* (London: Faber & Faber, 1995).
35 That said, there are new women dramatists that are adopting more experimental forms and style of work, as evidenced, for example, in Debbie Tucker Green's debut drama *Dirty Butterfly* (Soho Theatre, 2003) – though arguably there are relatively fewer 'experimentalists' than there are 'realists'.
36 This view has been widely supported by feminist theatre scholars such as Sue-Ellen Case and Jill Dolan. Exceptionally, there have been a minority of bourgeois feminist scholars that have sought to reclaim or to 'rehabilitate' realism. See, for example, Sheila Stowell's 'Rehabilitating realism', *Journal of Dramatic Theory and Criticism*, 6.2 (Spring 1992): 81–8.
37 A much commented on example of this is Sarah Kane's debut play *Blasted* (1995), which opens with a recognizable, realistic hotel bedroom setting that gets torn apart and becomes increasingly unreal or surreal, in terms of the play's composition, aesthetics and performance style.

6
Predicting the Past: Histories and Futures in the Work of Women Directors

Aoife Monks

> Theatre – is as a cultural activity deeply involved with memory and haunted by repetition.[1]

Theatre has a peculiar effect on time. In its practice of making the past present, of embodying texts and performance styles hundreds of years old, theatre introduces fragments of history into the present: constructing, creating and imagining pasts that are distinctively theatrical. The urge to represent the past on stage inevitably has ideological consequences, as Susan Bennett argues, 'often what is perceived as "lost" is reasserted by its cultural representation'.[2] Theatrical re-inventions of history are an intervention into the present, offering a critical relationship with contemporary issues through the lens of the past. Moreover, these theatrical representations of history implicitly offer a vision of the future, suggesting ways in which spectators might rethink concepts of progress, transformation and the *status quo*. In order to think forward, theatre practitioners frequently look back.

Theatre's ability to evoke the past in ways that signify to the present, explains its attraction for feminist practitioners and critics. Re-presenting the past has been key to a feminist engagement with the future: by offering a sense of multiple pasts, by activating the silent voices of history, feminist directors, writers and critics have suggested new possibilities for change, for the: 'social transformation of gender relations'.[3] Theatre's ability to make the past immediate is a means for feminist practitioners to offer an imaginary future that can work to critique present gendered hierarchies. However, theatre's tendency to recycle and reiterate its *own* past has been criticized for reproducing historical attitudes to gender. Feminist scholars have revealed how the male dominated canon of dramatic literature and performance traditions has contributed to the symbolic economies of gender oppression. A combined challenge to the status of the theatrical canon, and male-centred representations of the past, is central to a feminist engagement with futurity.

The representation of history on the stage must be a key way in which we understand theatre performance as 'feminist'. Nevertheless, this notion poses a series of problematic questions. For example, what if the past is represented on stage without offering the possibility of change in the future? Can this be classed as feminist theatre? On the other hand, even if a production does challenge gendered histories, can this work be considered feminist when the artists involved actively disavow any relationship with feminist theory or practice? In other words, can theatre be feminist when it's not intended as such? These questions have been raised by the emergence of some female directors in European and American theatre over the past 30 years, whose work has often been uneasily located within the canon of feminist theatre practice and criticism. This is particularly true of female directors in Britain and the United States since the 1980s, when the Thatcher and Reagan governments radically cut the funding for fringe theatre, traditionally the home of most feminist theatre practice.

These economic changes heralded a new breed of female directors who invaded the mainstream theatres of Britain and the United States and began to do work which was not consciously feminist and was more focused on representing the 'human' and the 'universal' in performance. While the presence of these women directors in the theatre broke new ground institutionally, their work did not necessarily challenge the artistic *status quo*, and many denied the significance of their gender to their work. Even so, I want to argue, that the work of these women has enormous value for feminist scholarship and frequently offers a critique of history and the canon, which has implications for how we might envision a feminist future. After all, as Alisa Solomon argues: 'a play's feminist meaning is to be found... in what happens to the audience.'[4]

The intention of the director is only one small part of how a theatre piece might be understood as feminist. Gender identity often occupies a deeply complex and conflicted terrain in the work of these women that produces fascinating, if not always entirely 'feminist', effects, through their engagement with the past. Nonetheless institutional 'location' is a factor in determining the reception of these effects and, in these terms, it is useful to compare what can be identified as three main categories of practice: the woman director in the institution, the woman director in the avant-garde, and the woman director in feminist practice. In each case, the use of casting in performances of canonical texts is a useful way to outline the different approaches to history in the work of these directors.

Different directions

Female directors working in the mainstream and in establishment institutions[5] (such as Katie Mitchell, Phyllida Lloyd and Garry Hynes), often support the status of the canon in their work, and do not challenge a male-centred version of history, or necessarily overtly engage with the question of gender. If gender is negotiated in their work, this exploration generally takes place within the auspices of the 'greatness' of the canon. For example, the directors of the recent plethora of cross-dressed performances in these institutional spaces often resist the idea that gender or the canon might be disrupted by such a casting choice. This resistance could be seen in the way Helena Kaut-Howson, the director of a cross-dressed *King Lear*, argued that: 'we cast Kathryn Hunter because I believe the part is about old age and not about gender. It should be available to women and men.'[6] While cross-dressed performances in this mode are often validated through the rhetoric of theatrical tradition (as with the cross-dressed productions at the Globe Theatre), female cross-dressing is frequently seen (and often dismissed) as 'gender-blind casting', akin to the colour-blind casting practices now commonly used on the mainstream stage. This view leaves the gender meanings within the plays themselves unchallenged, and positions cross-dressing as a means for female actors to attain the greatness of male actors on the stage. Cross-casting becomes an equal rights issue that erases the question of gender and does not work to critique the status of the text or to disrupt the image of history. Yet, at times there can be a clash between the representation of gender in these women's productions, and the conservative institutional context in which they work. This clash produces intriguing and controversial incarnations of gender on the stage, and places pressure on the ideological underpinnings of the humanist bent of these institutions.

Female directors (such as Ariane Mnouchkine, Ann Bogart and Pina Bausch) working in the avant-garde, offer deconstructions of gender in their work that are a by-product of a larger project of formalist experimentation. Here, the implications for gender in the use of deconstructive techniques – such as the use of male impersonation and non-naturalistic uses of costume, acting style and language – tend to go unnoticed by critics who overlook such implications in favour of an emphasis on the aesthetic qualities of the work, and a more generalized conception of 'politics'. An elision of the question of gender could be seen in Ann Bogart's claim that American artists should: 'be accomplices in the collective act of memory [and] engage in social criticism and attempt to rediscover an identification with historical continuity'.[7] While Bogart's work has been deeply enriched by its critical approach to American history, her argument does not acknowledge that 'history' is not a unified mass, but a multiple series of perspectives and narratives, all of which are inflected with the question of gender. This 'disappearance' of gender in Bogart's engagement with history, could also be seen in the critical reception of her use of cross-dressing, when she cast the actress Kelly Maurer as Andy Warhol in her production *Culture of Desire* (1995), which was based on Dante's *Inferno*. Except to note the ways in which the cross-casting supported Maurer's characterization of Warhol as 'feminine', critics by-passed the gender implications of the casting, in order to talk about the 'real' politics of Bogart's critique of consumerism. This response is typical in the reception of avant-garde practice by women directors, where questions of gender are often side-lined in favour of the supposedly gender neutral critique of mediation, consumerism or post-modernity. However, the tools of deconstruction and historicization used by directors such as Bogart do often produce critical representations of gender, albeit without necessarily offering the possibility of the transformation of gender hierarchies, and this work is of value to feminist scholarship.

Finally, feminist theatre directors and artists (such as Split Britches, Holly Hughes and Bobby Baker) are often situated in a similar economic and theatrical context to that of the women working in the avant-garde, but their work overtly focuses on the theatrical and social strategies required to challenge representational codes of gender. Their work often blurs the distinctions between the role of the director, writer and performer, and they frequently work within a collective structure (such as the WOW café in New York), in which democratic collaborative processes replace the hierarchies of traditional theatre practices.[8] Here, casting and cross-dressing are used to disrupt canonical texts in order to reconstruct gender relations. *Belle Reprieve* (1991), the collaboration between the feminist company Split Britches, and the male drag company Bloolips, was a case in point. This cross-cast reworking of *A Streetcar Named Desire*, exposed the heteronormative codes that underpinned the representational strategies of the text. Through the use of song, dance, parody and excerpts from the text, the performers made the queer dynamics of identity available to the audience through: 'an exploration of the relationship between narrative and social conformity, between the yoked constraints of naturalism and naturalized notions of gender' (Solomon, *Redressing The Canon*, p.158).

The work of feminist practitioners such as Split Britches has rightfully been 'canonized' in feminist criticism. However, while much of the work of women working as directors outside of this context might not be nominally feminist, it should still be situated within the growing and diverse body of theatre work that challenges the histories of gender representation and thereby: 'condition[s] our expectations for the future'.[9] In these terms, the work of directors Elizabeth LeCompte (with the Wooster Group in New York[10]) and Deborah Warner (with Fiona Shaw in Britain)[11] are of particular note. Neither director overtly raises the question of gender in their work, nor do they describe their work as feminist (especially Warner, who actively rejects the term). However, their work is a complex response to the ways in which history is central to the representation of gender. As prime examples of mainstream theatre practice (Warner) and avant-garde experimentalism (LeCompte), the work of these women demonstrates the ways in which gender has been critiqued and deconstructed within a non-feminist framework. I want then, to look at Warner's cross-dressed production of Shakespeare's *Richard II* in order to argue that her contradictory engagement with the idea of the 'universal' in performance revealed the masculinist bias of representations of monarchy on the stage. I will then go on to examine how LeCompte's treatment of Arthur Miller's *The Crucible* revealed the ways in which gender is not only imitated, but formed, through the traditions of theatrical representation.

Deborah Warner's *Richard II*

Deborah Warner claims that gender is irrelevant to her work, arguing that it is: 'less significant than the fact that she doesn't have a university background',[12] and she aims to reveal the essential 'human' resonances and truths within theatre texts. Yet her productions have contained deeply ambiguous (and highly praised) representations of masculinity and femininity. This ambiguity has primarily been channelled through the body of the actress Fiona Shaw, frequently described by critics as 'androgynous', who has exerted a deconstructive effect on the portraits of femininity in canonical texts such as *Electra* at the RSC (1988) and *Hedda Gabler* at The Abbey Theatre (1990). As a result, the work Warner and Shaw have undertaken together has often been underscored by tensions between the politics of identity and the claims of the authority of the writer. These tensions emerged most clearly in Warner and Shaw's *Richard II* at the Royal National Theatre in London in June 1995, with Shaw in the title role. Even before the production opened, the critics were in uproar at the prospect of Fiona Shaw playing Richard. This negative response raises the question as to why it was that the highly praised ambiguity she had brought to the performance of female roles was now deemed unacceptable in her portrait of a deposed male king? This reaction was especially puzzling given the long-established tradition of playing Richard as 'feminine' or rather, 'effeminate'. However, the critics' vitriolic response revealed that the past is gendered in history plays, with much of their opposition focusing on the inadequacies of Shaw's gender in her representation of monarchy. While they attacked her ability to play 'male', with one critic arguing that: 'Shaw doesn't have enough maleness to play Peter Pan,' the critics simultaneously criticized her for being inadequately feminine, with one critic describing her as: 'a lean, angular woman with a sharp jutting jaw [who] is not particularly attractive.'[13] In contrast to her positive reviews for portraying women, in which her ambiguity and complexity was praised, Shaw as Richard was then criticized for her inadequate femininity as an actor – or as a woman – and at the same time denounced for being 'a stereotypical girlie' (Koenig, 'Theatre reviews', p.10).

The contradictions in the critics' response, point to the cultural struggle around the production of Shakespeare's history plays, which have: 'done more to shape popular conceptions of English history than the work of any professional historian'.[14] Warner's production revealed the implicit gendering of the representation of history in the performance of Shakespeare plays. The effect of her casting choice operated in stark contrast to that produced through the medium of the male actor's body, such as when Laurence Olivier and Kenneth Branagh's film versions of *Henry V* fed the 'public longing for narratives of strong male heroes who embody national prowess' (Howard and Rackin, *Engendering A Nation*, p.9). By changing the sex of the actor's body, Warner had challenged how the past might be imagined and, by extension, how monarchy and Englishness were configured on the stage. While W. B. Worthen argues: 'the body becomes the vehicle for the transmission of Shakespeare,'[15] this production revealed that it is the *gendered* body that 'transmits' Shakespeare.

Nevertheless, in the publicity surrounding the production, Warner rejected the notion that she cast Shaw as Richard for feminist purposes, saying: 'It just struck me that on a very simple plane he is somewhat feminine.'[16] Similarly, Shaw viewed the role of Richard as enabling her to represent a state of universality which female roles did not give her in the theatre: 'what is fantastic, as a woman, is being allowed to play with the existential contradictions of the universe' (cited in Armistead, 'Kingdom', p.8). Rather than Shaw and Warner trying to address the problematic notion of Shakespeare's universality, they viewed cross-dressing as a means to access that universality, not usually available to women. Gender did therefore play into their decision to cast Shaw, but a critique of the gender imbalance within the authority of Shakespeare performance did not. As is more generally true of cross-dressing in mainstream productions, their approach to the question of gender positioned it as one of 'equal opportunities' rather than as a challenge to the representational system that has maintained those inequalities.

This tension between an acknowledgement and an elision of gender also manifested itself within the use of design and performance style in the production. Warner's show operated in an uneasy half-light between an overall realism in its approach to design and acting style, and the stylized, abstracted figure of Shaw's Richard. Shaw was wrapped in bandages from her shoulders to her ankles, visible beneath her medieval-style costume. The bandages denaturalized and stylized the image of her body and worked in contrast to the historical realism of the costuming of the other actors on the stage. Moreover, while most of the actors performed in the mode of psychological realism, Shaw's performance was visceral, hyperactive and non-naturalistic, with much emphasis on her verbal and physical virtuosity. Her Richard was placed in poses that ritually displayed the king's body, and she moved from this three-dimensional portraiture into occasional moments of frenetic activity, such as jumping on to John of Gaunt's deathbed in a tantrum and starting a pillow fight. Only during the prison scenes at the end of the play was Shaw's body shown in naturalistic repose, operating perhaps as a comment on Richard's loss of theatricality after the loss of his crown and, notably, these were the scenes that the critics praised in reviews of the production. Crucially, Shaw was also the only cross-dressed figure on the stage, so that overall her presence continually disrupted the consistency of the realist field of the show and challenged Warner's otherwise psychological and 'faithful' approach to the text.

Warner's production, then, constructed a feminized, theatrical and deeply unstable past world embodied by Shaw, which was idealized by the production as an alternative to the masculine history embodied by

David Threlfall as Bolingbroke. Ironically, however, by casting Shaw in the role, the production could also be said to have maintained and reasserted many of the stereotypes around gender, essentializing both Richard's 'femininity' and Shaw's 'femaleness' on stage. The fact that Shaw played a male character that has been traditionally seen *as* feminine, or even effeminate, problematized Shaw's usual critical engagement with gender. Her performances of Electra and Hedda had offered complex portraits of intelligent women struggling against the restrictions of their environment, and her virtuosity and theatrical power had worked to counteract the disempowerment of these characters. Yet, as Irving Wardle pointed out, in her role as Richard: 'much of her performance conforms to the womanish stereotype: irrationally obstinate, unable to make up her mind, forever retreating from public issues into personal relationships.'[17] In short, Shaw's interrogation of female identity when playing *women* contrasted with her much more conventional portrait of 'femininity' – when playing a *man*. While Shaw claimed that playing a man gave her access to universal human truths, in fact she performed the most stereotypically feminine role of her career. Warner and Shaw ignored the fact that our notions of universality have been predicated upon a canon of literature and theatre that has been male-centred. The idea that Shaw could attain universality had mistakenly relied on the presumption that universality was not gendered.

However, the history and England that Warner offered through Shaw's performance still challenged the gendered dynamics of the representation of the past. Warner's production expressed a sense of nostalgia for the medieval moment which was shown through the reverence for ritual, the choral laments sung by female singers, and the evident loss even Bolingbroke felt for the destruction of Richard's world. Peter Holland saw this nostalgia as: 'a reactionary mood'[18], arguing that the production hearkened back to the glory days of the English past. However, it was the *kind* of past for which Warner's production expressed such longing that constituted the most radical component of her production. She constructed an image of medieval England which was ambiguously gendered, and which was ruled by a figure that combined masculine, feminine and childlike traits. Invoking a world in which gendered and sexual lines were blurred, Warner's production demanded sympathy for the ambiguous qualities of Shaw's Richard, and positioned the audience to regret the loss of his world. Warner re-imagined an already imaginary past, and by casting Shaw as Richard, she revealed the image of history as masculine to have been imaginary all along.

Warner's nostalgia for Richard's feminine and theatrical moment is particularly powerful if we take account of Susan Bennett's definition of nostalgia as: 'the inflicted territory where claims for authenticity... are staged' (*Performing Nostalgia*, p.7). Performing a history play at the Royal National Theatre with a sense of nostalgia for a past centring on a visceral, denaturalized and feminine figure has implications for how the present and future of the nation are envisioned. The sense of yearning that was so central to the production's logic, encouraged the audience to long for the ambiguities and femininity of Richard in their own world, to re-imagine their present as well as their past, and suggested an alternative future to that proposed by the more conventional masculinist vision of history on the Shakespearean stage. Even while Shaw's cross-dressed performance constructed stereotypical images of femininity on the stage, Warner's production reconfigured the intertwined relationship of Englishness, monarchy and masculinity in the representation of history. As Carol Chillington Rutter argues: 'perhaps gender did matter after all. Perhaps only a woman playing the king could estrange the role sufficiently for this demystification to happen and to permit a British audience to consider what a very odd idea a "king" is.'[19]

The Wooster Group's *LSD* (*...Just the High Points...*)

As with Warner's production at the National Theatre, the context in which Elizabeth LeCompte works with the Wooster Group informs how the relationship between embodiment, the canon and the representation of history and gender is received in her productions. The Wooster Group's 30-year reputation for experimentation has bestowed the company with a high status in the international community of spectators for avant-garde performance. The company's history of controversy in their early years has ironically made their work safe in the latter part of their career by creating the expectation of shock, which contains the impact of some of their riskier experiments. As a result, the ways in which gender is deconstructed in their performance practice has often been overlooked in the critical reception of their work. However, I would argue that the question of gendered embodiment and its relationship with history is one of the central touchstones of their productions. LeCompte frames, mediates and distorts gendered bodies in her productions, and ultimately shows gender to be the product of representation itself. LeCompte's politics are relativist and uncertain, and her interrogation of the authority of texts is key to her challenge to the orthodoxies of gender representation. The performer's 'presence' is both central to the company's practice and is simultaneously undermined and fragmented by their use of video monitors, microphones and live cameras on the stage. Similarly, play-texts are fragmented and dispersed, competing with the meanings produced by technology, sound design and performance, rather than being positioned as a central source of meaning.

LeCompte's description of her political approach to representation also suggests a gendered dynamic to how she makes theatre: 'That's one of my things about male writers of the 50s, ... their ability to pinpoint right and wrong. Miller is so clear about it. I can't be clear. As a woman of the 60s, 70s, 80s, I can't be clear. I don't know who the enemies are. I don't know if there are enemies.'[20] By entering into a dialogue with the authority of these writers, LeCompte challenges the gendered *status quo* of the canon. At the same time, her interrogation of these texts rarely functions as an outright rejection of their values. Rather, the Wooster Group have situated themselves within the legacies of American experimental theatre (such as the work of Thornton Wilder or Arthur Miller), while at the same time consistently interrogating their own ethical and political relationship with this work. As a result, their productions place the spectator in a deeply ambivalent relationship with the images and meanings they produce on their stage.

The link between theatre history and the question of gender was particularly prominent in The Wooster Group's production *LSD* in 1984, which interspersed the trial scenes from Arthur Miller's *The Crucible* with a reconstruction of a rehearsal of Miller's play by the company while on acid; readings from the work of Beat poets; the reconstruction of a public debate between Timothy Leary, the LSD guru, and Gordon Liddy; a film of Ron Vawter, one of the company's actors, in Miami; and a series of dances. These elements were positioned in unframed relation to each other, and formed a kind of theatrical collage in which the spectator was forced to read between the various performance modes in order to create meaning. It was in these in-between spaces of performance that LeCompte located her critique of the gendered dynamics of canonical representation. By placing Miller's text in circulation with a series of other social, cultural and theatrical texts, LeCompte challenged the text as the 'centre' of meaning in a theatrical production and revealed it to have been a product of its historical time and place. Furthermore, she exposed the underlying hierarchy of gender representation upon which Miller's own social critique rests.

The Crucible is a play that interrogates the consequences of a social morality founded on extremes of good and evil. It draws a comparison between the McCarthy trials and the Salem witch trials of 1692. Miller's play makes a case for the necessarily ambiguous and complex nature of goodness through the figure of the flawed hero, John Proctor, whose moral struggle is at the centre of the politics of the play. However, in order to render Procter a hero, Miller relies on the vilification of the female characters in the play, especially Abigail. Miller describes Abigail as: 'seventeen ... a strikingly beautiful girl, an orphan with an endless capacity for dissembling',[21] while Mercy, her co-conspirator is described as: 'a fat, sly, merciless girl of eighteen' (Miller, *The Crucible*, p.24). The manipulative nature of the girls who 'cry witch', is offset by the moral strength of Proctor in the play, and Miller achieved this effect by universalizing Procter's individualistic qualities while simultaneously generalizing the unstable and repugnant characteristics of the female characters. In *LSD*, The Wooster Group questioned the characterization of Abigail as a 'mad ... murderous bitch', by implicitly linking Abigail's story with that of Timothy Leary, the LSD guru, whose own subversive attempts to challenge the social *status quo* were carried out through dubious means.[22]

When the Wooster Group performed excerpts of *The Crucible*, the female performers were dressed in seventeenth-century costume, whereas the male actors were costumed in clothes from the 1950s, 1960s and 1980s, embracing the period when Miller wrote his play and the contemporary moment of the audience. The audience were therefore positioned to identify with the familiar and individualized qualities of the male actors' costumes. On the other hand, the women were distanced from audience identification by wearing the clothes of the historical moment in which the play is set. The historical and conformist nature of the clothing generalized their appearance and drew more attention to the 'image' of their bodies than to their interior or individualized subjectivity. Through the use of costume, then, LeCompte heightened the processes of identification already at work in Miller's play. The gendered dynamics of Miller's play were further upheld through the use of technology. The male performers spoke through microphones but the female performers did not, struggling to project their 'live' voices over the amplified male voices. LeCompte noted that: 'the live performer has to shout very loud and give an immense emotional output to equal a whisper on a microphone. So a lot of the performance played off huge emotional vocal outputs against very tiny verbal outputs into the mic.... The women got the costumes, the men got the mics' (LeCompte, cited in Aronson, 'The Wooster Group's *LSD*', p.72).

We might ask how these theatrical strategies challenged the gendered dynamics of Miller's play? By linking the female performers with the 'image' of history through costume design, and the male actors with the textual authority of the play through vocal amplification, the production might be seen as maintaining the traditions of gender representation in theatre performance. However, historical distance was *literalized* by this use of costume and technology and LeCompte undermined the authority of the spoken text by exaggerating its qualities and by rendering the status of the text equivalent to the other performance modes in operation in the production.

LeCompte's use of amplification actively foregrounded the way that female voices are obscured and marginalized by the play text. Taken together, the hyperbolic qualities of casting, costume and sound offered the possibility of subversion by placing the audience at a distance from the representation of gender in Miller's play, so as to allow them to recognize and critique the artificiality of this mode of representation. *LSD* underlined the fact that the canon and the traditions of performance often rest on a problematic hierarchy of gendered representation, and LeCompte pushed these stereotypes to such an extreme that they became denaturalized and historicized.

The critical impact of exaggeration was further maintained by the fact that while the male actors/characters were associated with the text, they spoke their lines at breakneck speed, garbling the meanings of the words. Miller's text therefore became a rhythmic structure that was rendered theatrically equivalent to the dances and the flickering television screens on stage. As a result, the audience could not garner 'more' or 'greater' meanings from the text than they could from the other elements of the production – unlike a more conventional production of *The Crucible*, there was no hierarchy of signs in operation on the stage. As Phillip Auslander points out, within the Wooster Group's own work the 'image is far more central . . . than the verbal text . . . their images frequently undermine, ironise or overwhelm texts'.[23] For example, the use of the dance pieces in *LSD* did not serve to illustrate Miller's play, elaborate on character or extend the narrative, but retained an autonomous function – the dance contained its own meanings and status on stage, rather than being put to use in order to 'serve' the greater meanings of the text. The traditional hierarchies of theatrical representation – with the authority/author of the text at the top – were reconfigured by rehabilitating the value and power of the image and by counterbalancing and undermining the authority of language. As a result, the challenge to the status of language simultaneously became a challenge to the traditional hierarchies of gendered representation and while men retained the traditional associations with language, and women with image, the status and significance of these positions was critiqued and refigured in performance.

However, the wider social authority of the text was revealed in the lawsuit from Arthur Miller forbidding the Wooster Group's use of material from his play in their production. In 1983, Miller met with the company and voiced three concerns: 'first, the audience might think LeCompte's interpretation a parody. Secondly, the audience might believe the piece was a performance of the entire play and not just excerpts. And thirdly, Miller feared that these performances might exclude a "first-class", i.e. Broadway production' (Savran, 'Arthur Miller and *The Crucible*', p.100). As David Savran notes, the irony of the fact that the Wooster Group later had to close their production, due to Miller's lawsuit, was that the play-text itself questioned the use of 'arbitrary, inflexible and overzealous authority' ('Arthur Miller and *The Crucible*', p.109).

Much has been written about Miller's action and about what this revealed around authorship and the role of the text in performance.[24] However, what is missing from this debate is the question of gender. It would be a mistake to read the opposition between Miller and LeCompte in biological terms, but it is possible to analyse it as a semiotic battle between masculine certainty and a relativist and subversive feminine sensibility. Miller's presumption that LeCompte's production would not be 'first class' belies a larger attitude to the work of women in the theatre, assuming that they cannot maintain the standards set by 'great' (mainstream) theatre, and seeing her work as a parody rather than recognizing the more complex pastiche she created. Miller mistook a complex acknowledgement of the cultural status of his text for an outright rejection of its values. The fact that the play was incorporated into the Wooster Group's work demonstrated their attraction to its cultural status, and to its radical intention, and while they were critical of his representations of gender, they just as powerfully interrogated their own attitudes to gender and representation more generally.

LSD simultaneously functioned as a performance of a male canonical writer's play, and a counteracting narrative by a female avant-garde director. Yet, the performance did not offer any alternative images of gender to those supplied by Miller's text. The spectator was unable to assume a superior or controlling position of knowledge over the relationship between the texts, visuals and bodies in action on the stage. Unlike Miller's text, which offered a new set of certainties in place of those that he critiqued, LeCompte's production placed his work in a process of circulation with other cultural, political and performance texts, offering a performance of uncertainty, collision and contradiction. Through the disjunction between Miller's work and the cultural texts from the intervening period between the time of his play and the moment of the Wooster Group's performance, LeCompte revealed that Miller's desire for radical social change had been undergirded by an unconscious ideology of gender hierarchies that his radicalism could not escape. The logic of foregrounding this fact meant that LeCompte also had to acknowledge her inability to transcend her own historical context by refusing to offer alternative images of gender in place of those supplied by Miller. History – both theatrical and social – was shown to be formative and inescapable in how we see and act gender identity on stage, challenging the sense of the artist as 'outside' of culture, or superior to the ideologies that they critique.

Nevertheless, by pushing these images to such extremes, the company did offer a sense of the subversive possibilities of theatrical and social performance. In a Butlerian sense, the hyperbolic nature of the Wooster Group's performance offered the possibility of doing gender 'wrong' and denaturalizing its seeming hegemonic inevitability. Gender stereotyping was resisted through its wholesale, exaggerated endorsement in this production, a strategy that always runs the risk of maintaining what it seeks to undermine. LeCompte showed gender identity to be inextricably caught up within the traditions of representation and the canon, but while the historicity of gender was revealed and foregrounded in *LSD*, the future remained difficult to imagine, bound up as it was in the endless and repetitive cycles of repressive traditions of representation.

Conclusion

The work of women directors, frequently reveals the gendered assumptions underpinning the narratives of history and the position of the canon in performance. In Deborah Warner's *Richard II*, the context in which Warner worked, and her avowal of humanist principles, created a productive tension within her use of cross-dressing, which unintentionally revealed the gendered assumptions underlying the concept of the 'universal.' Her production controversially longed for a feminine, ambiguous and theatrical future that emerged from the contradictory nature of the performance itself. On the other hand, Elizabeth LeCompte's work with The Wooster Group launched a critique of the gender systems underpinning the canon without offering a sense of possible alternatives, being unable to envision a transformed future for gender representation.

In both cases, the questions of history and the canon were at the centre of the controversial ways in which gender was represented in these productions. Neither of these directors publicly describes themselves or their work as feminist, and their work has not been read as such by critics. However, the interrogation of the past in this work offered a critique of the ways in which theatrical representations of women can limit the possibilities for lived experiences of gender, and challenged (even if unintentionally, on the part of Warner) the view of the male-dominated theatrical canon as 'universal'. There was an implicit plea in the work of these directors for a future which challenges the legacies of the past and for a recognition that we are bound by history, while retaining the desire for change and transformation. As Neil Jarman argues: 'It is the desires and aspirations of the present that shape our views of the past, while at the same time those present aspirations are partly formed by our understanding of our past.'[25]

Notes

1 Marvin Carlson, *The Haunted Stage: The Theatre As Memory Machine* (Ann Arbor: University of Michigan Press, 2003), p.11.
2 Susan Bennett, *Performing Nostalgia: Shifting Shakespeare And The Contemporary Past* (London and New York: Routledge, 1996), p.3.
3 Judith Butler, *Undoing Gender* (New York and London: Routledge, 2004) p.205.
4 Alisa Solomon, *Redressing The Canon: Essays On Theater And Gender* (New York and London: Routledge, 1997), p.14.
5 While distinctions between 'mainstream' and 'fringe' or 'avant-garde' theatre have been contested in recent years, here the distinction is useful in order to delineate the ways in which female directors describe their work, and are positioned by their context in relation to the question of gender.
6 Dan Glaister, 'Welcome Queen Lear', *Guardian*, 23 June 1997: p.1.
7 Ann Bogart, 'Stepping Out Of Inertia', *The Drama Review*, 27 (1983): 26–8, p.27.
8 Elaine Aston, *An Introduction To Feminism And Theatre* (London and New York: Routledge, 1995) p.59.
9 Geraldine Harris, 'Introduction to Part Two', in Lizbeth Goodman with Jane De Gay, eds, *The Routledge Reader In Gender And Performance* (London and New York: Routledge, 1998), pp.55–9, p.56.
10 The Wooster Group were founded in 1975 as part of the Performance Group, with Elizabeth LeCompte at the helm. The company are renowned for the collaborative methods they use to produce their work. In 1980, the company named themselves The Wooster Group and in 1981, they performed their first production under their new name, *Route 1&9 (The Last*

Act) which was hugely controversial for its juxtaposing of Thornton Wilder's *Our Town* with blackface minstrel routines and a porn film. They produced *LSD (... Just The High Points ...)* (1984); *North Atlantic* (1984); *Frank Dell's The Temptation Of St Antony* (1987); *Brace Up!*(1991); *The Emperor Jones* (1993); *Fish Story* (1994); *The Hairy Ape* (1995); *House/Lights* (1999); *To You, The Birdie (Phedre)* (2001); and their most recent production, *Poor Theatre: A Series Of Simulacra* (2004).

11 Deborah Warner was born in Oxfordshire, England, in 1959, and founded Kick theatre company in 1980, which she ran for six years. In 1988, she was appointed resident director at the RSC and directed Fiona Shaw, in Sophocles's *Electra. Electra* began a collaborative relationship which had a major impact on both their careers. Warner and Shaw went on to stage *The Good Person Of Sichuan* (1989); *Hedda Gabler* (1990); *Footfalls* (1994); *Richard II* (1995); *The Waste Land* (1995–7); *Medea* (2000); and *The Powerbook* (2002). Warner has recently developed a series of site-specific performances, and she recently won an Obie for her *Angel Project* (2003) which was part of the Lincoln Centre festival in New York in 2003.

12 Claire Armistead, 'Women directors', *Women: A Cultural Review*, 5 (1994): 185–91, p.190.

13 Rhoda Koenig, 'Theatre reviews' *Independent*, 5 June 1995: 10; Richard Hornby, *'Richard II', The Hudson Review*, Winter (1996): 641–4, p.642.

14 Jean E. Howard and Phyllis Rackin, *Engendering A Nation: A Feminist Account Of Shakespeare's English Histories* (London and New York: Routledge 1997), p.10.

15 William, B. Worthen, *Shakespeare And The Authority Of Performance*, (Cambridge, New York, Melbourne: Cambridge University Press, 1997), p.99.

16 Claire Armistead, 'Kingdom under siege', *Guardian*, 31 May 1995: 8.

17 Irving Wardle, 'The woman who would be king', *Independent On Sunday*, 11 June 1995: 13.

18 Peter Holland, *English Shakespeares: Shakespeare On The English Stage In The 1990s* (Cambridge, New York, Melbourne: Cambridge University Press, 1997) p.247.

19 Carol Chillington Rutter, 'Fiona Shaw's Richard II: The girl as player-king as comic', *Shakespeare Quarterly*, 48 (1997): 314–24, p.323.

20 Le Compte cited in Arnold Aronson, 'The Wooster Group's *LSD (... Just The High Points ...)*', *The Drama Review: Choreography (And The Wooster Group's LSD)*, 29 (1985): 64–77, p.75.

21 Arthur Miller, *The Crucible, A Play In Four Acts* (Middlesex, New York, Victoria, Ontario, Auckland: Penguin Books, 1982), p.18.

22 David Savran, 'The Wooster Group, Arthur Miller and *The Crucible*', *The Drama Review: Choreography (and The Wooster Group's LSD)*, 29 (1985): 99–109, p.206.

23 Philip Auslander, *Presence And Resistance, Postmodernism And Cultural Politics In Contemporary American Performance* (Michigan: University of Michigan Press, 1992) p.93.

24 See David Savran, 'The Wooster Group, Arthur Miller and The Crucible': 99–109; David Savran, *Breaking The Rules: The Wooster Group* (New York: Theatre Communications Group, 1988); Arnold Aronson, 'The Wooster Group's *LSD (... Just The High Points ...)*', *The Drama Review: Choreography (and The Wooster Group's LSD)*, 29 (1985): 64–77; Philip Auslander, 'Task and vision: Willem Dafoe in *LSD*', *The Drama Review: Choreography (And The*

Wooster Group's LSD), 29 (1985): 96–8; Auslander, *Presence And Resistance*; Gerald Rabkin, 'Is There a Text on This Stage? Theatre, Authorship, Interpretation', in Rebecca Schneider and Gabrielle Cody, eds, *Re:direction: A Theoretical And Practical Guide*, (London and New York: Routledge/TDR, 2002), pp.319–32.

25 Neil Jarman, *Material Conflicts: Parades And Visual Displays In Northern Ireland* (Oxford and New York: Berg, 1997), p.5.

7

The Screens of Time: Feminist Memories and Hopes

Sue-Ellen Case

At first, it seemed that feminist futures were to be found in feminist pasts. The beginning swells of the second wave deliriously projected images of a feminist past onto the bifurcated screen of time, reimagining matriarchies, amazons and goddesses along with the secret lives, the so-called untold histories of those few women history had managed to recognize. Feminist historians, literary critics, and artists filled pages and stages with these mythical creatures whose lust, power, love and accomplishment stunned and seduced the gray, tired so-called 'male' histories. The story telling of the grandmothers and their ghostly performances in the passing flesh, displaced the notion of history with that of cultural memory. However termed, these projections logged images and events on to the screen of time, establishing an account of cultural capital for the present movement. The feminist present and the feminist future were animated by these figures, these actions, these settings deemed as remembered. In one sense, many were imagined utopias, hopes for the future, embedded in the past. The bipolar screen of time, the past and the future was a form of collective dreaming through temporal tropes.

Yet, growing up alongside and within these temporal twins, a slipstream of time, a wormhole of time, imagined through the notion of palimpsests complicated the strict division of past, present and future. In the artifact itself, in the present performance, one could perceive the layering of memories, image painted upon image, the erasures, the re-engravings, the gestural remains and even the traces of the future. Although these traces are located within a circumscribed material such as a canvas, or a square where performances had taken place, the strict chronology of those figures could not be ascertained. No development or progression could be marked. Uncertainty haunts the image, where erasures have occurred, or where images overcome the traces beneath them. Synchronicities appear there and it is difficult to separate out projections of the future from memories of the past.

Semiotic slippage put time on a banana peel, prat-falling on to the stage of representation. These time slippages can be hopeful and bright, in the tradition of early feminist imaginings, but they can also be dark reminders of exclusion. In specific, the very representation of 'woman' both set things in motion and determined their endpoint. As Teresa de Lauretis put it, in her 1984 book, *Alice Doesn't* the representation of 'woman' serves as 'the source of the drive to represent and its ultimate, unattainable goal'.[1] 'Woman', then, a cultural myth made to stand in for women's experiences, catalyses both the past and the future – the beginning and the end of representation. Gender codes are, in fact, time codes, complicit with the structuring of the desire and the dream into partitioned, policed zones.

On a happier note, in performance studies, Joseph Roach, in his book *Cities of the Dead,* launched the notion of 'surrogation', in which contemporary performances could be seen as embodying the past in present performances. Now one could view 'expressive movements as mnemonic reserves, including patterned movements made and remembered by bodies, residual movements retained implicitly in images or words (or in the silences between them), and imaginary movements dreamed in minds not prior to language but constitutive of it'.[2] The present is perforated by the past in this new form of understanding embodied histories. Even improvisation, the form meant to signal the present in its momentary invention, is a process of remembering and forgetting in its gesticulations. Somatic memories propel the body through space, making the performer a time traveller. Performances are both portals into an unknown future and kinesthetic and oral reserves of the past.[3]

The closet of time

If de Lauretis marked an exclusion as temporal ground and future, queer scholars began to identify other forms of exclusion that they could liberate – semiotically bringing past works 'out of the closet' into the brighter present and future of queer epistemologies. Returning to the notorious old queen – of the novel *Remembrance of Things Past* – who opened the twentieth century with the discovery that memory could be embedded in the body, Proust could now be queered. Proust had narrated how eating the Madeleine, or stumbling on the pavement were kinesthetic sites of memory that could invoke the past in a more complete and vivid fashion than any voluntary memory could perform. These episodes were kinesthetic theatres of memory, in fact, replete with characters and a detailed *mise-en-scène*. But his own voluntary writing could also be brought out of the closet through new readings of its involuntary inscriptions. Proust's epic novel bound his queer body to an intricate interweaving of cultural and personal memory, performed through transgender identifications, inextricably binding sensual and sexual escapades with spectacles of class difference and cultural production. *The Remembrance of Things Past* mapped out the critical topography of the twentieth century at its inception. However, as Freud and Lacan have pointed out, each thing that is remembered is another forgotten. Proust's closet of memory, formed by cultural restrictions against his transgender identifications and his sexual proclivities displaced his homosexualized liaisons and longings on to phobic images of queers such as Albertine, who is portrayed as performing seemingly condemnatory acts, such as cavorting with another woman in front of her mother's portrait; while his intimate portraits of social circles were bound tightly to the 'right' bloodlines. Rereading phobic images as desires, the past of Proust and his world can be liberated into the present queer moment as a monument to the trans-world of the early twentieth century.

Such semiotic slippage can be made to serve its own master, as D. A. Miller wittily and elegantly configures in his book *Bringing Out Roland Barthes*. Miller asserts that 'With increasing visibility after *S/Z*, Barthes is engaged in the ambiguously twinned projects of at once sublimating gay content and undoing the sublimation in the practice of what he himself refers to – in the case of Proust's inversion – as *form*.'[4] By form, here Miller means Barthes's own semiosis. Miller organizes a way to read Barthes's play of sign and referent, by referring to his own experiences with the discourse of the sub-culture, thereby recognizing how the specifically gay codes are deployed through slippage, and other devices designed to create an esoteric homosexual semiosis. Thus, Miller's knowledge of gay districts in Tokyo allows him to read, say, Barthes's *The Empire of Signs*, to reveal a different register of referents that mark a gay Tokyo and offers another, more seductive way to read how Barthes may have been looking into the eyes he discusses there.

Such 'outing' processes, slipping through transgender inversions and tortuous, complex projects of sublimation and inscription, have brought to the stage a space for utopic projections of future possibilities into past dreams. Split Britches and Bloolips, in their production of *Belle Reprieve*,[5] 'out' Tennessee Williams through transgender casting and the development of the sexual subtext within *A Streetcar Named Desire*. The laughter and the cheap jokes demonstrate how the present possibilities for sociosexual identifications and relations are free from the repressive coding in the text. Yet it is precisely that past which shapes the joke. They perform a present and promise a future in their songs that are 'outside' the Williams text, addressed directly to an audience bound to the history of the theatre of repression and despair. When the suffering, tortured character of Blanche Dubois is transformed into a 'critical fairy' (a postmodern drag queen) whose rival is a woman playing Stanley Kowalski (Peggy Shaw), the sub-cultural queer semiosis overcomes, through slippage and gender inversion, the fixed, determinist codes of the 1940s. Moreover, as Miller could find the gay Tokyo beyond Barthes's *Empire of Signs*, the *mise-en-scène* of *Belle Reprieve* literally flattens out Williams's seemingly realist referent into painted drops that signify psychic processes of identification. The light bulb Blanche seeks to cover with her veil of illusion is bared on a painted flat, isolated in its signification, no longer part of the 'realistic' setting of a bedroom. The play slips, semiotically, into the symbolic and imaginary registers of sexual and gendered identifications, surrogating, as Roach would have it, the past, while also stepping out of its hold. Time becomes, as it is in dreams, Freud argues, simply an imagined structure of the negotiations of power and desire. Memory is embodied as Peggy Shaw's Stanley cries out the mantra of sado-masochistic pain/pleasure in the by-now famous 'STELLLLA'! Yet her self-reflexive play on the passage safely ossifies that past, while still invoking it in the present, and also promises release from its determinism. The theatrical past and the historical fixing of the author are everywhere present and abandoned.

Queer scholar José Esteban Muñoz identifies what he terms 'the future in the present' outside of the heteronormative sense of it. Muñoz notes that heterosexual culture 'depends on a notion of the future: as the song goes, 'the children are our future'.[6] Seeking to establish a different tradition for a future in the present in 'cultures of sexual dissidence', Muñoz argues for an 'anticipatory illumination of a queer world' ('The Future in the Present', p.93). In a sense, the performance of *Belle Reprieve* can be understood through this critical lens as performing that anticipatory gesture in Williams's play.

The subject of time

Although feminists began to have their way with time, reflexively participating in its screening effects and, like Alice, sliding happily across its semiotic fields and wriggling through its wormholes, its deterministic dynamic still took hold. While the shape of time could be fashioned, its power to produce meaning could not. The agent, or the subject, became a kind of side-effect, rather than operator of the time machine. Those earlier feminists who were animating the past with matriarchs and mermaids were doing it alongside the feminist project of recuperating a subject position for women. Some feminists took the motto 'the personal is the political' of the activist movement into their cultural work, seeking to abandon the then-termed patriarchal voice and procedures of the objective historian or critic. The personal 'I' figured woman's empowered standpoint, in locating a site of negotiation among issues of class, sexual practices, racialized codes and, of course, gender, while also retaining the power to alter them. However, the great cleansing of the 'I' through attacks against essentialism encouraged another form of slipstreaming, or wormholing. Applications of psychoanalytic theory both created the interrogation of the subject and swapped the personal for the cultural. Displacement through multiple objects, or Lacan's familiar *'objet petite à'* creates an ego that is an archive of the approaching and discarded objects of identification. Thus, the 'I' is an archive of the past and a site of possible identificatory streaming.

But looking back from the now-more-varied readings in critical theory, we can see that even the so-called objective tradition of writing theory was only one form of philosophical inquiry, which had been remembered in order to displace another. Another notorious homosexual, Ludwig Wittgenstein, whose *Tractatus Logico-Philosophicus* was lionized by, of all things, logical positivism, considers the operations of the subject in this painful passage marked by sublimation and inscription in *Culture and Value*, or *Vermischte Bemerkungun: 'Nobody can truthfully say of himself that he is filth.* Because if I do say it, though it can be true in a sense, this is not the truth by which I myself can be penetrated: otherwise I should either have to go mad or change myself.'[7] One might say, given the status of the notion of 'shame' in Sedgwick's theories of queer performativity,[8] that Wittgenstein launched the queer performative voice in his consideration of Kierkegaard and Christianity in 1937. But with this important difference: Wittgenstein's 'self' was only partially the performative one, completely constructed or hailed by the dominant ideology that would 'shame' him. Instead, he remains truly split in his shaming of himself, for the subject who shames cannot be the one who is shamed – for Wittgenstein, that would be a logical and grammatical impossibility, implying madness. Unlike Althusser, for whom the subject is interpellated, 'hailed' by ideology, always already in an object position,[9]

Wittgenstein implies that in order to mobilize the system of shaming, or ideology, the subject must retain its power to initiate. If the subject has been displaced by a functionalist or structuralist kind of objecthood, for Wittgenstein, it can logically and grammatically only inhabit that object status by retaining its bipolar relationship to the status of the subject. Interestingly, for Wittgenstein the only way the subject could be understood as pure object would be as one who goes mad, which means that one cannot operate within linguistic or logical systems.

Sarah Kane's play *4:48 Psychosis* scripts the mad roles that the 'I' as subject and object may play – particularly through the process of shaming.[10] In this play, psychosis is both a psychological and a social condition, rather than an aspect of a character, for the play has no characters in the traditional sense of the term. Psychosis is the state of mind and the mind of state that fractures social affiliations, as well as any lasting cohesion of the subject position. The personal and the political play across the interior and social field of the play. The remains of character are occlusions of shame and shaming, stopping and being stopped. They act either as restrictions, as in the deployment of therapy-ese, or as refusals, stopping desire, change and identity formation on the part of the subject/object patient position. Much of the agency in the play is stoppage at the site of interpellation: 'I can't eat'/'I can't sleep'/'I can't think'/'I cannot love'/'I cannot fuck'/'I cannot be alone'/'I cannot be with others' and 'I do not want to live' (*4:48 Psychosis*, pp.206–7). Pronouns substitute for nouns in much of the text, with purposefully indistinct referents, so the 'I' hangs (itself) as a position in the psychotic social space. Take, for instance, the following use of the 'I':

> I gassed the Jews, I killed the Kurds, I bombed the Arabs, I fucked small children while they begged for mercy, the killing fields are mine... and when I die I'm going to be reincarnated as your child only fifty times worse and as mad as all fuck I'm going to make your life a living fucking hell I REFUSE I REFUSE I REFUSE LOOK AWAY FROM ME.
>
> (Kane, *4.48 Psychosis*, p.227)

This is both the personal 'I' of early feminism, with its desire and its social positioning marked, but it is also the 'I' of the state. It is the speaking of self through the past of its violent oppressions and of its future in reincarnations of violence. But the refusal stops the determinism of both time and repetition. Although only a stoppage, it is powerful in halting the run of revenge.

With nothing like a characterological position, a subject in reserve articulates its actions in a string of unmoored repetitions: 'Slash wring punch burn flicker dab float'. These strings are followed by groups of three nouns. One group is particularly vibrant for the condition of the theater as locked to the social: 'Victim. Perpetrator. Bystander' (Kane, *Psychosis*, p.231). Here is a key to the dark scene – one cannot pretend not to witness, not to see, and thus not to partake. Sometimes collecting itself into the 'I' of the storm, Kane installs the subject of time as psychotic, mad in Wittgenstein's terms. The social past of that subject, in killing Kurds and bombing Arabs, is its future, in reincarnation. Her madness is in opting out of the system altogether – the violence of her escape is like the violence of the system itself – the madness of a war zone, the loss of the subject position that could control the verb, the action, wrested from her by the authority of doctors and the state.

Kane's 'anticipatory' gestures, as Muñoz would term them, reside in what could be termed a 'negative utopia'. Borrowing from Ernst Bloch, one could find a wish, a 'residue' 'that is not fulfilled and made banal through fulfillment...'.[11] Or as Adorno, in conversation with Bloch, imagines, 'the contradiction between the evident possibility of fulfillment and the just as evident impossibility of fulfillment' which compels the spectators to 'identify themselves with this impossibility and to make this impossibility their own affair' (Bloch, *Utopian Function*, p.4). The longing for change inscribed in the play, seeking semiotic and social liberation through the reception of its performance is its negative utopia. The future must be activated in a present so determined by its past. Of course, ascribing the resolution of this contradiction to Kane's suicide would release the spectators from this obligation – deny them the utopic promise in the piece. Melancholy would delimit the response, invoking Kane's own death as the final resolution of this anticipatory gesture.

Wrapped in the American flag, Holly Hughes acts out the address Kane impounds in the performance of her play, *Preaching to the Perverted.* Hughes scripts her journey up the steps of the Supreme Court in 1998 to hear their decision on the appeals the 'NEA four' made against censorship of the arts in the case *Finley v. NEA*. Four artists, with queer performance materials, had their grants revoked by NEA (the National Endowment for the Arts) in 1990 for offending 'standards of decency'. Staging the dark forces of state power as a comedy, in contrast to Kane's tragedy, Hughes appropriates the flag that waved against her and the others who were 'stripped' of their NEA funding. The Supreme Court justices appear as nine yellow rubber ducks who hear the arguments.

Hughes:

> But the obvious question doesn't get asked:
> How does the government decide someone is decent?
> ....
> Perhaps they used my mother's method:
> They check your underwear.
> First, it has to exist; no drip-dry advocates need apply
> It must be all natural fibers, preferably white
> No holes, stains, sagging waistband, safety pins!
> Maybe a simple pattern but no slogans please!
> No 'I can't believe I ate the whole thing'
> And it's got to be appropriate:
> No boxer shorts on women
> No lacey thongs on men.[12]

Richard Meyer observes that 'rather than countering these attacks by defending her work as respectable or decent, Hughes insists upon the symbolic power of the unrespectable and indecent' (Meyer, 'Have you Heard', p.552). Hughes mobilizes the juridical past to argue her case, in her own constituted queer performance court, by soliciting an identification with the perverted and their cause against the puritanical prudery in the Supreme Court decision. She invokes, then, a virtual court that could reverse this decision, changing the future for the queer arts within the state.

Hughes's performance piece was more prescient, more pregnant with the future than we could imagine. Now, some few years later, the Bush administration has been re-elected on the grounds of so-called decency in its civil struggle against gay marriage. Conflating 'homeland security' with the exclusive right to marriage by heterosexuals, the Bush administration hopes to mount the steps to the Supreme Court in order to pass a new amendment to the federal constitution, denying the rights of marriage to any but the heterosexual couple. The future was definitely embedded in the past, in this performance by Hughes. Only the laughter, the cutting courage of Hughes's humour can securely wrap the queer body in state protection and privilege. Hughes installs an agency in the citizen subject increasingly at risk.

Hughes discharges what Ernst Bloch imagined as the '*undischarged, underdeveloped, in short, utopian*...' gesture.[13] The future and the past are fully installed in her performance present. Indeed, much promise has been laid at the feet of performance art, from Josette Féral's early essay,[14] to

Peggy Phelan's description of its ontology,[15] to Diana Taylor's utopic transdisciplinary claims for it in her recent book on Archive and Repertoire.[16] Most recently, some feminist scholars, in considering the work of Hughes and other performance artists, have located utopic effects in the conditions of performance itself, installing future promise in present conditions. Live performance, argues Jill Dolan, can produce the 'feel' of a utopic future/present in a collective setting. Dolan terms this the 'utopian performative', that offers, in its practice, a sense of 'what utopia would feel like rather than how it would be organized. It works at the level of sensibility, by which I mean an affective code…'.[17] Dolan regards the space of performance as a 'space apart' in which 'Staying together in that space can be a time of shared subjectivity' ('Performance, utopia, and the "utopian performative"', p.468). Dolan ultimately moves to a discussion of the 'Messianic' in Deborah Margolin's performance work, finding what she terms 'messianic moments' in works 'that herald the arrival of a new and better world' ('Performance, utopia, and the "utopian performative"', p.476).

Rather than a content, then, that can be semiotically liberated, or a negative dialectic embedded in the cultural artifact, Dolan would offer the experience of certain performances, the affective charge they produce in those present, as the discharge of the hope, the coming of the messianic moment that 'heralds' a better future. Dolan posits the collective moment that has been confounded by state practices, corporate immersions and technological isolation in the conditions of live performance, where its utopic potential may be realized by certain kinds of performance.

But returning to Bloch, who was attacked by other Marxists for a kind of 'messianic' subtext in his 'principles of hope', we can see just why he configures the utopic or the present as specifically 'unfulfilled'. Unlike Dolan, or perhaps even Augusto Boal, who would have people imagine the better world and realize its image or affect, Bloch wants to hold it within a kind of 'negative dialectic'. When utopia is 'cast into a picture', as he puts it 'there is a reification of ephemeral and non-ephemeral tendencies'. For Bloch, the utopic resides in the 'it-should-not-be' of the longing for a 'coming in order'. 'The function of utopia', for Bloch, 'is a critique of what is present' (Bloch, *Utopian Function*, pp.11–12). To actually discharge it would be to reify it.

Elfriede Jelinek, who won the Nobel Prize for Literature in 2004, inscribed such negative subjects-of-promise in her play *Illness or Modern Women (Krankheit oder Moderne Frauen)*.[18] In this play, the men speak only in hyperbolic platitudes of health, strength, efficiency, success, competition, sports and so on. Dr Heathcliffe is both a dentist and a gynaecologist. His nurse, Emily (Brontë) is a lesbian vampire. Emily falls in love with a patient, Camilla (Le Fanu), a housewife/mother and wife of a tax collector. Camilla dies in childbirth in the gynaecological chair as Emily sucks her blood. Brought back to life by vampiric means, Camilla returns Emily's affection and celebrates her new status of 'undead'. As the women bond through blood, the men continue their mutual backslapping, congratulating themselves on their various successes. Their narcissism and sexism blind them to the vampiric satiety the women achieve with one another. Act II opens on a 1950s style bedroom with twin beds, actually twin coffins, in which Camilla and Emily lounge as Emily writes on her portable typewriter. A negative utopia is staged with great irony, as the women find fulfilment in their negative status, dripping with blood, haunting graveyards and drinking from medical beakers of blood. Finally, the women are able to merge into a double-headed single creature that feasts on all flesh and blood. The women have successfully removed themselves from social functions as well as any romantic attachment to nature through their undead condition. The lesbian vampires are both deconstructive figures evacuating by their presence the dominant social logic.

However, their very fulfilment, the 'fleshing out' of their promise also brings, as Bloch would predict, their demise. The double-headed creature is destroyed. Its representational fulfilment brings its demise. To come out of the closet, in this play, is to depreciate future promise. And yet, the fact that Jelinek, a feminist author despised in her home country of Austria for her critical works, could be awarded the Nobel Prize for literature does seem somewhat fulfilling. Perhaps, as Bloch would have, it, the contradiction is what promises hope.

At this point in history, it seems difficult to imagine any kind of a promising or feminist future. The re-election of the Bush regime in the United States seems to promise nothing but suffering to people and to the environment. The rights and privileges of citizenship are being restricted by immigration laws, the lawless incarceration of terrorist suspects, moral censorship of the arts, and homophobic amendments to the constitution. In spite of Jill Dolan's enthusiasm for the realm of radical performance, it seems more and more divorced from social conditions. In fact, performances focused on singularly human concerns seem more and more untied from the crisis in the environment, with increasing numbers of animals becoming extinct, whales and other fish beaching themselves and dying in record numbers, the arctic poles melting, and the future of a sustainable environment more and more unlikely as water reserves begin to disappear.

With crises of domination of nature and other humans forming on every front, the title of a performance piece by the lesbian artist Reno seems evocative: *Rebel Without a Pause*. In the past, queer scholarship revealed the relations between formations of citizenship and minoritarian sexual practices. Sarah Schulman's *My American History*,[19] the Parker/Rouseau/Sommer/Yeager anthology entitled *Nationalisms and Sexualities*[20] and David Evans's *Sexual Citizenship*,[21] all traced, in various ways, the formation of sexual citizenship as an effect of national and juridical processes. But after the traumatic pressure the events of 9/11 unleashed on national and juridical identificatory processes, the lesbian performance artist Reno crafted a piece in which she reversed the equation: the events of 9/11 are not made to reveal the citizen status of the lesbian, but the lesbian performer reveals the militaristic isolationist violence of the state.[22] If, in the past, the practice was to reveal the status of one who identified with a sexual minority within the processes of citizenship and legislation, this performance stages the lesbian performer as one who speaks about the nation as a citizen – looking out from that vantage rather than playing out issues surrounding her unique identity. This point was made to me by David Román who invited Reno to perform this work at the University of South Carolina precisely because he wanted to foreground the importance of gay/lesbian performers dedicating their work to a rebellion against national processes.

It seems ironic to conclude an article on feminist futures, once located in an autonomous movement, a separatist culture, an identitarian sub-culture, with a consideration of the nation-state. This is not to suggest that the nation-state is the effective form of organization, as in the modernist, bourgeois tradition; nor would I like to pose a non-gendered kind of politic that returns to the unmarked status of 'citizen'. The promise I can hold here, in *Rebel Without A Pause*, is in the residue, the contradiction between rebellion and take-over, the negative dialectic of aggression and expenditure, which makes the audience responsible for change. This residue exceeds the gesture of response and responsibility in the rebel as subject, tied to activist performers in Seattle, say, with giant puppets performing against the WTO, or those garbed as tomatoes and other vegetables demonstrating against the genetic alteration of 'frankenfoods', or the marching masks of Bush in the anti-war marches, or the masked women at sewing machines in front of the Acropolis, staging the sweatshop production of athletic clothing during the Olympics. As old as Medea (to slipstream back into the past), the woman terrorist who poisoned the state and murdered its future, who escaped by chariot into the utopic heavens, while the rebellion brims, stirred in large cauldrons by witches and vampires, visible, like those secret signs of the underground movements, agitating, remaining behind masks, in the dark holes of aporias, never really to actualize, be recognized, punished or tamed.

Notes

1. Teresa de Lauretis, *Alice Doesn't: Feminism, Semiotics, Cinema* (Bloomington: Indiana University Press, 1984), p.13.
2. Joseph Roach, *Cities of the Dead: Circum-Atlantic Performance* (New York: Columbia University Press, 1996), p.26.
3. Joseph Roach, 'Deep Skin: Reconstructing Congo Square', in Harry J. Elam, Jr. and David Krasern, eds, *African American Performance and Theater History: A Critical Reader* (Oxford and New York: Oxford University Press, 2001), pp.101–13.
4. D. A. Miller, *Bringing Out Roland Barthes* (Berkeley: University of California Press, 1992), p.27.
5. See Bette Bourne, Paul Shaw, Peggy Shaw and Lois Weaver, '*Belle Reprieve*', in Sue-Ellen Case, ed., *Split Britches: Lesbian Practice/Feminist Performance* (London and New York: Routledge, 1996), pp.149–83.
6. José Esteban Muñoz, 'The Future in the Present: Sexual Avant-Gardes and the Performance of Utopia', in Donald Pease and Robyn Weigman, eds, *The Future of American Studies* (Durham, NC, and London: Duke University Press, 2002), pp.77–93.
7. Ludwig Wittgenstein, *Culture and Value*, G. H. Von Wright, ed., trans. Peter Winch (Cambridge, MA, and Oxford: Blackwell, 1980), p.32e.
8. See Eve Kosofsky Sedgwick, *Touching Feeling: Affect, Pedagogy, Performativity* (Durham, NC, and London: Duke University Press, 2003).
9. Louis Althusser, 'Ideology and Ideological State Apparatuses (Notes Towards an Investigation)', in *Lenin and Philosophy and Other Essays*, trans. Ben Brewster (New York: Monthly Review Press, 1971), pp.127–88.
10. Sarah Kane, *4.48 Psychosis, in Sarah Kane: Complete Plays* (London: Methuen Drama, 2001), pp.203–45.
11. Ernst Bloch, *The Utopian Function of Art and Literature: Selected Essays*, trans. Jack Zipes and Frank Mecklenburg (Cambridge, MA: MIT Press, 1988), p.2.
12. Quoted in Richard Meyer, 'Have you heard the one about the lesbian who goes to the Supreme Court: Holly Hughes and the case against censorship', *Theatre Journal*, 52 (2000): 543–52, p.551.
13. Ernst Bloch, *The Principle of Hope*, trans. Neville Plaice, Stephen Plaice and Paul Knight (Oxford: Blackwell, 1985), p.102.
14. Josette Féral, 'Performance and theatricality: The subject demystified', *Modern Drama*, 25 (1983): 170–81.
15. Peggy Phelan, 'The Ontology of Performance: Representation without Reproduction', in *Unmarked: The Politics of Performance* (London and New York: Routledge, 1993), pp.146–66.
16. Diana Taylor, *Archive and the Repertoire: Performing Cultural Memory in the Americas* (Durham, NC: Duke University Press, 2003).

17 Jill Dolan, 'Performance, utopia, and the "utopian performative"', *Theatre Journal*, 53 (2001): 455–79, p.460.
18 Elfriede Jelinek, *Krankheit oder Moderne Frauen* (Köln: Prometh Verlag, 1987).
19 Sarah Schulman, *My American History: Lesbian and Gay Life During the Reagan/Bush Years* (New York: Routledge, 1994).
20 Parker, Rousseau et al., eds, *Nationalisms and Sexualities* (New York: Routledge, 1992).
21 David T. Evans, *Sexual Citizenship: The Material Construction of Sexualities* (London and New York: Routledge, 1993).
22 See Reno, 'Reno: Rebel without a pause: Unrestrained reflections on September 11th', http://www.citizenreno.com/indexrebel.html, viewed 6 December 2004.

8
Africa Lives On in We: Histories and Futures of Black Women Artists

SuAndi

'The keepers' is my term for, and my way of introducing, the artistic voice of Black women that gives expression to familial and cultural ancestry. In Africa we call it *Sankofa*: to look backwards in order to look forwards. Native Americans call it *Neshkinukat*: one who has relatives. Both suggest the 'keeping' of community, history and tradition, a fabric of ancestral links threaded through contemporary living. Sometimes lost, always changing, these ancestral threads are kept alive particularly by women through their traditions of oral history.

Juliet

She likes the emptiness
Bringing space onto the platform stage
Of where she remembers.
Toning lights, so as to not
Tint the memory
Of the profit making productions
That so many laboured on.
Can't change what is,
Has always been,
Nowhere else to go,
How, what, could we,
We have no choice but to follow

In the steps of mother, father, brother, sister
Both grandparents
And sometimes a great granddad.

She moves through the darkness of what was.
And hears again the voices.
The complaints
Asides:
The women's smuttiness of the young unskilled labourer
Still a boy, working as a man.

Here, north and south came together
In sweat encased tombs of toil
And outside the towers
Belched out a haze of things will get better
But they didn't.

In the end, they all died
Some in their time
Some in their prime
Before they should have.
Some left clasping cheap gold chains
To watch the idle days ahead
And from their seat in the audience
They timed the slow but determined
Crumbling.

The last time the whistle blew,
They heard it.
The last gate to clank close
They witness it.
The last chimney to darken the sky,
They saw it.
Until the steel of silence fell
All around them
And only the odd streetlight glowed
A memory
She stands now in its hazy embrace
Shadowed by missing walls
People
Machines.
Bringing up yesterday
To tell us how she

Daughter of the south
Growing in the north
Remembers.

(© SuAndi 2005)

Sistah-in-Arts

Juliet Ellis is a Black[1] artist who lives and produces Live Art in Manchester in the north of England.[2] She has the loudest laugh on earth and I consider her to be a Sistah-in-Arts. When we get the chance to meet and chat it is inevitable that we talk about clothes, men (well boys), food, body shapes and the arts. We discuss where we find our work sits or how we try to find a place in the arts when a dominant few are determined to keep us out.

Juliet's work is anything but linear and because of this she sometimes frowns when I say that we create work that tells stories, our stories. Juliet isn't sure that her shows work in this way. Her latest piece, *The Meeting Place*, is part of a major body of work, an *Autobiography in Five Chapters*. In an exploration of the uncanny forces of dreams and memory, *The Meeting Place* seeks to challenge the boundaries of perception. An emotional landscape of human resistance is realized through a combination of film and live performance; Juliet's delivery of the piece is both unnerving and dark. The work tells those audiences with the imagination and the 'eyes to see' that Juliet's present self is also trapped in the industrial past of England's great city of steel, Sheffield. She doesn't dwell on racism. She doesn't want racism to be the focus of her work in an issue-based way, even though this is very much a part of her experience. Rather, *The Meeting Place* tells how it was for this particular Black woman, wandering the streets of a city whose pulse was tempered by the factories that soared above it.

She visits Sheffield in memories and dreams to gather up her family and to laugh, her voice echoes through the crumbling, derelict half-erect structures of the city's past. For some, this landscape offers ugly reminders of a collapsed economy that left men idle on street corners and mothers worrying about how to keep home and hearth together. It suggests hardship, the hardship of low wages, and too long a day's toil. But Juliet was a child in those days. She remembers the noise of the factories as they emptied in the evening. The musty odorous smell of the workers she passed as she played in the streets, and the magical intrigue the buildings held for her as she paused outside gates that restricted her entry. Maybe in another life she worked there, for truth to say, her voice has the power to soar above the clanging noise of any production line. Maybe.

Story-telling, self-exploring and performing

As with most writers, story-telling, or as my mother called it 'blatant lying', came naturally to me early on in life. For me, the everyday was too, too normal. In my naivety, I thought of my family as being especially boring and oh so normal! I longed for something more exotic, something that would make the other kids listen to me. In time, but only in time, I would realize that I had this already. I had it from my father's bloodline. I had Africa.

But before I made the link between me here and my family over there, I worked my way through (and out of) other possibilities. Shirley Bassey?[3] Too loud. When she sang her mouth opened as far as the racist images of Black mouths with thick lips. Eartha Kitt?[4] She was too sexy, not at all like the kittens I played with. Too much the oversexed image of Black women. I didn't feel I belonged to either, but what did I, this half-woman with a child's mind, know? So I invented a visiting cousin with skin so dark it shone like ebony. They had to be visiting England for all my Manchester cousins are mixed raced like myself and though we all have different skin tones they are without exception more brown than black. It would be many a long year before I realized that my linkage to Africa was more than the colour of my skin – that Africa is in my genetic memory: the way I walk, the carriage of my head, the need for ritual in my life and, most of all, the retelling of stories.

Juliet and I are different whilst at the same time we are very much the same.

Juliet and I are not alone. So many Black women artists are self-exploring in their performances. Telling stories of an immediate past that has resonance with hundreds of years ago and in doing so we are celebrating our position globally and locally as Black women.

Racism and celebration of self

When I began to lose myself in African-American literature it only served to open a deep wound of wanting for a 'steep stoop' on which I could sit shelling green beans as a long list of celebrities from the poet Langston Hughes to Sacthmo (Louis Armstrong)[5] and Malcolm X[6] dropped by. When it was these folk found the time to write, make music and change the world didn't occur to me at the time, and as I had no sense of the geography of the United States I didn't stop to figure out how they could be dropping by so many brownstone residences. I am not trying to say that the writers were fibbing, as my mother would have said, but what did trouble me was the absence of hardship, the absence of the white man with his racist laws and brutal oppression. It would take me years to understand that even with a life trespassed by race laws Black families get up, plan their day and see the time through from sunrise to sunset.

Racism does not position us in society. We achieve out of rebellion. We push back boundaries and obstacles that are built by racism. We must be creative in order to live above the limitations of racism. We have to tell our stories so that globally our children will know them, and so that the ancestors know that even though we have drifted so very far away from our first village of Africa we have never forgotten the griot rituals.[7]

My own writing took many years to develop from its initial base of the poet telling stories, those sometimes, blatant lies that my mother would scold me for, into a literature archive of my time. I try to allow my writing to evolve from where I am in order to give me directions towards where I aim to be in the future. But in all of this, it is important that the writer concentrate on the history she lives and is not 'burdened' by the race issue.

Borrowing Kobena Mercer's[8] phrase 'the burden of representation', Stuart Hall has drawn attention to and questioned the artist's need to represent the Black community.[9] The problem, as Hall perceives it, is how to get away from the idea that a Black artist has to speak for a whole race. 'If the phrase ["burden of representation"] has any meaning at all', Hall explains, 'we now have the burden of struggling on two fronts – to gain access and visibility, and then, within that, for different meaning of the black experience, which can never be unitary.'[10] Hall's argument is echoed in the performance and writing experiences of award-winning playwright and actor Kwame Kwei-Armah.[11] Kwei-Armah's success with his debut play *Elmina's Kitchen* at London's National Theatre (2003), which attracts predominantly white, middle-class audiences, could be seen as gaining the kind of visibility that Hall argues for, though at the same time, risks being (mis)read as the 'quintessentially black' play. 'Somehow we're so tribal', writes Kwei-Armah, 'that they'll just go "Oh that one's for the black audiences, so we can miss that one"'.[12] Kwei-Armah insists, however, that he is not writing primarily for Black people but about the truth that he knows: 'I do not attempt to speak as a representative voice of Black communities.'[13] He explains:

> As a writer I simply want to tell damn good stories. As a Black-British writer, I'm interested in creating narratives that celebrate my cultural inheritance while challenging negative stereotypes.
>
> (Interview, *Elmina's Kitchen*)

For Kwei-Armah this necessarily involves 'reclaiming Africa' and acknowledging his West Indian roots at the same time as refusing the history of slavery, 'an illegitimate act' that he doesn't wish 'to carry . . . into [his] every day life', his future (Interview, *Elmina's Kitchen*). Similarly, I see that many of my friends and colleagues locate their identities in their cultural heritage, as in Jamaica, Trinidad, India and so on, and so far across the globe it evidences that to describe any people of colour as a *minority* is the biggest joke of all.

Speaking for myself, I am quite comfortable being Black British. As the daughter of a Liverpool Mother and Nigerian father I meander between cultural tags including Mixed Raced, Black British, Nigerian and simply Black. It is the Black that I am most comfortable with which might, well, seem ironic because my most successful writing has been based around my white mother and through her 'voice' the exploration of racism on family life.[14]

Returning for a moment to my early story-telling that my mother called lies. At first I simply told stories. Stories I believed filled in the gaps which I found in the family home. But was it as crude as all that? Were my stories possibly the weaving of tales, weaving interwoven with odd facts, imaginary people who replaced certain folk who were missing or who I would have preferred gone? Was I, still an aspiring teenager, simply putting a gloss on the worn everyday painting of life? Maybe the early acting out of characters and roles, and the evolution of all this into multi-layered characters of myself were carved out of the memorized images of the woman who had scolded me?

In fictionalizing the past if not the present, I did not realize, for a very long time, that I was planting roots for the future. I was putting down foundations, sturdy ones from which pathways could be walked, to avoid the potholes that I had tripped over and sometimes fallen into. I did all of this without knowledge. Knowledge came later from the constructive criticism of others.

I would begin to learn that there were many others like me. Thousands of writers, women writers, Black women writers who were on a sojourn from the past to the future. Writing 'her-stories' to give voice to those who had been silenced. Voices screaming to get out. Humble voices, strong brave ones, voices without the accent of class, mouths full of Ebonics. All of them women's voices, creative and enduring in their labour.

'Getting Black right. Right?'

The Arts for me are a school of learning as well as an arena of resistance. One of the many things I have learnt and resisted through the Arts is the way Black Arts get positioned and represented – especially by funding bodies. I am not alone in this view. The Manchester-based playwright Sonia Hughes, for instance, writes:

> I keep thinking I don't write from a Caribbean/British perspective, my work just is. But time and again it keeps biting me on the arse. I have stories to tell and the stories come from who I am and what I know and what I'm trying to find out about who I am and what I know. I try to illuminate the silences we have between ourselves.
>
> I do write from a Caribbean/British perspective, I was just scared to acknowledge it in case it was thought of as folksy, community – not art. But it is.[15]

Hughes expresses the fear shared by many Black artists regardless of the genre of their work. This *putting down* of Black art has been a battleground for all of us working out of the United Kingdom. The ever-changing arena in which the arts infrastructure places our work has only served to further frustrate our creative endeavours. In the last 20 years we Black artists have been named as 'Multicultural', 'Ethnic Minorities', for the briefest period of time 'Black', then elevated (or so they claimed) to 'Cultural Diversity' and then hammered down again to the anagram of 'BME' (Black Minority Ethnic).

At the National Black Arts Festival held in the United States at Atlanta Georgia in 2004, the 'Post-Black' Visual Arts Panel held a round-table discussion about whether African-American artists have a special responsibility to deal with race in their work. Whilst African-American artists (artists of colour) of the 1920s made work that was rooted in their African past, some new-generation artists no longer feel the necessity to work ancestry in this way. As a concept 'Post-Black' is meant as an elastic term that can take in Black artists whose desire is to be freed from the 'burden' of representing race issues. However, different views on this obtain. Michael Harris, Black British playwright and 'Post-Black' panellist in Atlanta, argued the need for an artist to have a 'cultural rootedness' that he (sic) can build on in whatever way he wants.

In making his argument, Harris quoted from Karla Holloway's *Codes of Conduct* (1995):

> 'Act your age, not your color.' I turned only slightly; but just enough to make certain I had not imagined these words – that they had actually been spoken by someone and had not slipped away from a distant childhood memory.[16]

Harris explained:

> Even in Black communities, Black was a negative signifier; an indicator of discredited conduct, ethics, and appearance. We were inscribed and circumscribed by our color – by how we looked! There simply was no way out because Blackness, unlike Jewishness or Irishness, is primarily visual (though certain speech is thought to symbolize Blackness as well).[17]

On the other hand, there are those artists who rebel against the negative of 'Blackness'. Consider, for example, this statement from Kojo Griffin

> I am not trying to ignore my cultural heritage. I would like to expand what a Black artist can deal with . . . I want to deal with commonality in human behaviour. Besides. What can I add to the canon of black art that hasn't already been said?[18]

Yet a Black cultural heritage is hard 'to ignore'. As Hughes points out (see earlier), as much as she tries not to write from a Caribbean/British perspective it just keeps filling up the page.

There are those who (critically) question the concentration of energy on the past coming from many different Black communities. It is often linked to the why-don't-you-leave-slavery-where-it-happened-in-history view. I have some sympathy for this view, although not, I would stress, on the subject of slavery itself. But, as more and more of the arts and culture in Britain gets engrossed in 'yesteryear', I worry about how this takes on a (dangerous) nostalgia for the past: oral projects, celebrations or acknowledgements that link to sovereignty, war, discoveries and achievements, that have little to do with Black histories. We can't afford to leave our histories behind. The importance of telling our stories is not just the release of a creative flow. It is filling the voids of history otherwise tomorrow will be tainted and bloodied over and over again. As the UK-based African-American playwright Anita Franklin observes:

> One of the things I've seen over and over in Black theatre is the way in talking about our pleasures and heartaches we frequently are reminding someone out there of our accomplishments as a people. We as writers and performers sneak in history lessons, reminders and clues of who we are beneath the mask.[19]

Writing out of histories, that whilst oppressive also give us strength, and telling stories of our achievements against the odds, weaves stories of the past into the present. The 'keeping' of community blends the old and the new. We look backwards to look forwards. You can see this sometimes in our actions when, without even knowing it is happening, we replicate the actions of our grandmothers. We repeat, tell over, similar stories but in and for different times and places. We have knowledge passed on to us, handed down, that we may not have understood and yet are making ready to hand down and pass on again: ancestral pasts circulating in a living present, rooting and routing futures.

The Story of M

> In her mother's voice, Yvonne Brewster, Jamaican born actress, director, and founder of Talawa Theatre[20] company, quotes George Bernard Shaw: *'for unless you do your own acting; write your own plays, your theatre will be of no use. It will in fact vulgarise and degrade you.'*[21]

Ancestral lines, bloodlines, and telling stories all came together in my solo piece *The Story of M*. In this show I retell the life of a white woman, my mother, and her experience of raising her two Black children, my brother and I. I end with a few words about M as a final illustration of many of the threads in my discussion.

As a Black woman playing my white mother, I 'colour' the audience's view of M. The book is judged by the cover: the audience is convinced that they are witnessing the life of a Black woman. They fail to recognize that M's language, her whole persona is white. Women in the audience warm to her: as a single mother who marries in haste to give her son the security of a family and then, just as quickly, gets divorced to raise the two children on her own. M's resilience and caustic humour draw the audience in. But 'taken in' by the warmth and love of 'M', they are forced, close up, to see the many prejudices she faced – prejudices that they themselves may have been, or possibly still are, guilty of. Young, 'Black', pregnant, unmarried – all the very worst kinds of prejudices attach to M.

But M is not Black. And this audience confusion (of) with her identity is passed down to thinking about her children. How is it possible that a (white) woman, whose own upbringing was in the confines of a Catholic children's home, could educate her two children about their African heritage? M doesn't attempt this. Instead she instils in her children a sense of honour for their bloodline to Africa, a bloodline maintained through constant contact from the father:

> 'You see your father, well I can't stand the bastard.
> But he's an African – that means you're
> An African
> Never forget that'.
>
> (*The Story of M*, p.16)

In crossing racial lines M, a white working-class mother, is 'writing' her children's future histories into Africa. Moreover, it is this telling of stories that disputes and ridicules dominant and *fixed* ideas about race and identity: the media's myth, for instance, that children born out of mixed parentage are children tossed aside for adoption and fostering. A theatre audience knows and feels closer to 'home truths' of cross-racial kinship through the fictions of story-telling and performance of this kind of Black artistry, than anything the academic textbook purports to 'tell'.

> I am the keeper of my mother's story.
> Her strength of character forms the confidence in me,
> the daughter.
> My father's bloodline is my bloodline to Africa, links me
> here with my family over there.

I Feel My Presence

> I feel my presence
> And in my being
> Know there is no greater might than my own
> I stand like a child
> On the precipice of knowledge
> Nodding with the wisdom of grandmothers

I feel my presence
I linger for moments in a past so bloodied
That ghosts turn to carrion of human flesh to block my
progress in the future
My arms have embraced generations.
Hung from crosses of missionaries.
Impaled by civilisation
and still my soul has remained true
I feel my presence.
See my reflection
In many eyes turned blind to me
And know I am beautiful
Know I am Black
And rejoice that Africa lives on in me.

(© SuAndi 1999)

Notes

1 Black is used throughout as a political term for people of African, Asian and Caribbean heritage.
2 Juliet Ellis is Resident Artist at the Greenroom, Manchester 2004–5.
3 Born on 8 January 1937 in Cardiff's Tiger Bay, Shirley Bassey is best known for her unequalled three James Bond themes – 'Goldfinger', 'Diamonds Are Forever' and the rather forgettable 'Moonraker'. She became Dame Shirley Bassey in 2000.
4 Born in South Carolina on 17 January 1927, Eartha Kitt is ranked at 89 on VH1's 100 Greatest Women of Rock N Roll.
5 Louis Armstrong, the most famous jazz trumpeter of the twentieth century, was nicknamed Satchmo, an abbreviation of 'satchelmouth', a joke about the size of his mouth.
6 Born Malcolm Little in 1925, Omaha, Nebraska, Malcolm X became the most famous minister of the Nation of Islam after joining in 1952. In 1964, following a pilgrimage to Mecca, he converted to orthodox Islam, took the name El-Hajj Malik El-Shabazz and broke with the Black Muslims.
7 'In the annals of West African history, "griot" (pronounced gree-oh) was the honoured name bestowed on wise and knowledgeable storytellers entrusted with the pivotal task of documenting tribal histories and genealogies'. 'The urban griot', http://www.blackauthors.com/, viewed 3 May 2005.
8 Kobena Mercer writer and lecturer on the visual arts of the Black diaspora.
9 See Stuart Hall, 'New Ethnicities', in Houston A. Baker, Jr., Manthia Diawara and Ruth Lindenborg, eds, *Black Cultural Studies: A Reader* (Chicago and London: University of Chicago Press, 1996), pp.163–72.
10 From personal email correspondence with Stuart Hall, March 2005.
11 Kwame Kwei-Armah's *Elmina's Kitchen* (National Theatre, London, May 2003) won him the London Evening Standard Charles Wintour award for

'Most Promising Playwright' and the play was also nominated for the 2004 Laurence Olivier Theatre Awards for Best New Play of 2003. Touring regionally in 2005 at major urban city venues, with Kwei-Armah in the leading role, *Elmina's Kitchen* is also scheduled for a run in London's West End.

12 Kwame Kwei-Armah, interview with director Angus Jackson, in programme notes, *Elmina's Kitchen*, regional tour United Kingdom, 2005.

13 Kwame Kwei-Armah made this point while taking part in a discussion of his work in the *Today* programme broadcast on BBC Radio 4, 13 December 2004.

14 My mother's experiences of Liverpool family life from the 1940s through to the 1980s are dramatized in my solo performance, *The Story of M*, commissioned in 1994 by the Institute of Contemporary Art (ICA) and published in SuAndi, ed., *4 for More* (Manchester, Black Arts Alliance, 2002), pp.1–18.

15 Sonia Hughes, from personal communication with SuAndi. For further details of Hughes's writing and performing visit 'Slam poets – Information on UK performance Poets' http://www.flippedeye.net/slampoets/ukpoets.html, visited 3 May 2005.

16 Quoted by Harris, round-table discussion, National Black Arts Festival Atlanta, Georgia, 2004.

17 Harris, round-table discussion, National Black Arts Festival Atlanta, Georgia, 2004.

18 Kojo Griffin, round-table discussion, National Black Arts Festival Atlanta, Georgia, 2004. Griffin is arguably the highest profile contemporary visual artist living in Atlanta. See 'Kojo Griffin: Work', http://www.millerblockgallery.com/artist_pages/Kojo_Griffin/Kojo_Griffin.html visited 3 May 2005.

19 Anita Franklin, personal communication with SuAndi.

20 Talawa is the longest established Black Theatre Company in the United Kingdom, its first production was *The Black Jacobins* in 1986.

21 Yvonne Brewster, *The Undertaker's Daughter* (London: Black Amber Books: 2004), p.69.

9

The Politics of the Personal: Autobiography in Performance[1]

Dee Heddon

> Performance artists are often folks for whom 'the personal is political' remained a vital challenge, rather than a piece of seventies' kitsch or an excuse to pass off attending Twelve Steps groups and aerobics classes as contributions towards social change.[2]

The practice of performances that draw on personal, lived experience is coterminous with the history of 'Second Wave' Western feminism. As soon as the phrase 'the personal is political' was coined, women performers and artists began to turn to their own experiences, using these as primary resources. The context for the use of the personal in performance was primarily feminist. Thirty years later, though 'autobiographical performance' endures, the context within which this work is now made, shown and witnessed is (perhaps) vastly different.[3] Given this, it seems appropriate to reflect on the use of the 'personal' in performance and its relationship to the political. In particular, we would do well to bear in mind Gerry Harris's insight that 'not all of the personal is political in *exactly the same way* and to the same effect'.[4] The manoeuvres of poststructuralist feminism should further prompt us to ask, 'Which personal?' and 'Whose politics?'[5] However, to assume that any problematization of the slogan 'The Personal is Political' is a recent development is to neglect the always various articulations and deployments of this phrase.

The politics of the personal

Jennifer: There is nothing inherently feminist about changing your real desires – whether it's for a rakish guy or Mahnholo Blahniks. Feminism is about getting in touch with your true desires – whether it's your ambition or your sexuality or your maternity....I believe in a feminism that strengthens my connections to my own desires.

Katha: I am all for following one's star and embracing one's 'true desires' – but who arranges the constellations in one's personal firmament?....'You go girl!' is a good slogan. But it's not the only thing women need to hear. They also need to hear, from time to time, that old, infuriating, favorite saying of the hairy-legged ancients: the personal is political.[6]

Though the feminist movement of the late 1960s and early 1970s operated alongside other movements of that time, for Imelda Whelehan 'what made [it] distinct from [these] could be summed up in the slogan "the personal is political"'.[7] As Robin Morgan claimed, 'Women's liberation is the first radical movement to base its politics – in fact, create its politics – out of concrete personal experiences.'[8] Personal experiences typically equated with experiences located in the 'private' sphere, whilst their 'collective' recounting through consciousness-raising strategies was precisely intended to challenge the so-called 'private' status of these, politicizing them in the process.[9]

The slogan, 'The Personal is Political', has engendered various interpretations and applications, so much so that in Bonnie Zimmerman's estimation, even by 1975, 'by and large, members of the women's liberation movement had grown uneasy about personal politics' as the 'ambiguities and contradictions became evident'.[10] Lynn Segal meanwhile, writing in 1979, suggested that the slogan had become 'interpreted in a very vague way, as though it meant that whatever you do, your actions have political significance' (*Beyond Fragments*, p.179), an interpretation that led some women to feel that their personal actions and choices were being 'policed' for their implicit politics (Whelehan, *Modern Feminist Thought*, p.74). Partly in response to the assumption of the 'personal' being 'policed' by 'second-wave feminists', the slogan has received its most recent spin, or 'Third-Wave' redeployment.[11] In this context, the personal is redefined as political because it *is* personal, and the personal is (a) right. The politics at stake are those of 'individual freedoms' or 'individual desires'. Doing what you want, because you want it, is framed as politically empowering. Kristina Sheryl Wong's statement is indicative: "There is a new thing called 'third wave' feminism that will open the door so you can embrace politics by being who you are!"[12]

Feminism, however, continues to be a diverse practice, with multiple positions. Thus, so-called 'Third Wavers' Baumgardner and Richards caution readers that 'personal choices shouldn't be construed as the same as political action' (*Manifesta*, p.19). Most significant in relation to the practice of autobiographical performance, Kathy Pollitt critiques the deployment of 'the personal' by arguing that '"The personal is political" did not mean that personal testimony, impressions and feelings are all you need to make a political argument' (quoted in Baumgardner and Richards, *Manifesta*, p.20). Further, as I aim to show, the public revelation of personal experience is not in itself, necessarily or always, a political act.

The politics of 'autobiographical performance'

Since first seeing Bobby Baker's performance, *Drawing on a Mother's Experience* (1988), I have been zealous about the political potential afforded by women's performances that deliberately and explicitly draw on personal experience. My persistent argument has been that such works, dealing with the various 'matter' of lives, matter greatly. My body of evidence resides in a vast body of practice, stretching over 30 years:

> Carolee Schneemann, Rachel Rosenthal, Suzanne Lacy, Faith Wilding, Laurie Anderson, Linda Montano, Hannah Wilke, Bobby Baker, Lenora Champagne, Pamela Sneed, Annie Sprinkle, Holly Hughes, Lisa Kron, Mary Duffy, Joey Hateley, Leslie Hill, Helen Paris, Lois Weaver, Peggy Shaw, Reno, Deb Margolin, Robbie McCauley, Terry Wolverton, Marga Gomez, Ursula Martinez, Terry Galloway, Penny Arcade, Kim Ima, Denise Uyehara, Carmelita Tropicana, Tami Spry, Karen Finley, Linda Park-Fuller, Susan Lewis, Vanalyne Green, Carla Kirkwood, Nancy Buchanan, Julie Tolentino, Kate Bornstein, Linda Benglis, Nan Golden, Marty Pottinger, Yvonne Rainer, Sonia Knox, Martha Rosler, Maya Chowdhry, Meredith Monk...

The sheer extent of 'autobiographical performance' inevitably renders it diverse and varied; nevertheless I have always sought to assemble it under a broad banner of 'political'. Thus, Faith Wilding's *Waiting* (1971) revealed the passivity and non-agency of women (thereby taking agency in the process); Judy Chicago, Suzanne Lacy, Sandra Orgel and Aviva Rahmani's *Ablutions* (1972) revealed the extent, frequency and impact of everyday violence against women; Carolee Schneemann's *Interior Scroll* (1975) determinedly figured a female artist as subject and authority of her own work, speaking back to the male artist/authority presumed to know and categorize/judge her; Terry Wolverton's *In Silence the Secrets Turn to Lies/Secrets Shared Become Sacred Truth* (1979) spoke the unspeakable act of familial sexual abuse, transforming victim into survivor.

My engaged readings of autobiographical performances have been both enabled and inspired by critics and performers who recognize the immense potential of autobiographical 'interventions',[13] where autobiography might be an act of reclamation, reinvention, transformation or survival. Sidonie Smith, for example, acknowledges that 'Autobiographical practices become the occasions for the staging of identity, and autobiographical strategies for the staging of agency.'[14] Or, with a slight but notably different inflection: 'Autobiographical practices become occasions for restaging subjectivity and autobiographical strategies become occasions for the staging of resistance.'[15] bell hooks's understanding of the political act that is autobiography is similarly inspiring: 'Oppressed people resist by identifying themselves as subject, by defining their reality, shaping their new identity, naming their history, telling their story.'[16] Or Liz Stanley's impressive claim that auto/biographies are preoccupied with a 'literary and political re-shaping of language and thus consciousness'.[17] Or Kenneth Plummer, who understands the political effect of personal stories that narrate sexuality: 'These stories work their way into changing lives, communities and cultures. Through and through, sexual story telling is a political process.'[18] Or Bobby Baker, who admits to the act of '*making* stories up to *make* sense of the world',[19] and Lenora Champagne who understands 'that emotional experience is a means of exploring cultural positioning and defining one's place in the world.'[20] Or Elizabeth Bell who contends that '"Marginalised subjectivities", the catch-phrase for those denied subjecthood in traditional Western conceptions, move from margin to center (stage) in performance.'[21] Or Jeannie Forte, who argues that 'actual women speaking their personal experience create dissonance with their representation, Woman, throwing that fictional category into relief and question'.[22] Or Linda Park-Fuller who recognizes her tale of surviving breast cancer as 'an attempt to break out of the prescribed, marginalised role of "patient-victim", and exercise socio-political agency in the world. That exercise of agency, in turn, circles back to transform and constitute me as actor-agent – as *survivor*'.[23] Or performer Tami Spry, who claims performative autobiography as:

> a site of narrative authority, offering me the power to reclaim and rename my voice and body privately and in rehearsal, and then publicly in performance. The process enables me to speak the personally political in public, which has been liberating and excruciating, but always in some way enabling.[24]

Given that the most visible 'origins' of women's use of personal material in performance lies in the feminist movement of the early 1970s, the link between the personal and political seems self-evident from the start. Consciousness-raising was an informing part of feminist art-practice as Faith Wilding's reminiscences make clear:

> During the first six weeks we met in each other's houses for consciousness-raising, reading, discussion and sharing our first art pieces. Once, after an emotional consciousness-raising session about street harassment, [Judy] Chicago suggested we make a piece in response.... Never in our previous art education had we been asked to make work out of a real experience.[25]

It is easy to forget, in the midst of the overwhelming deployment of the seemingly 'personal experience' in contemporary mass media, that prior to the feminist movement of the 1970s the 'personal' remained firmly private. In a 'pre-confessional' era, simply placing the 'personal' literally centre stage was in itself a radically political act, not to mention courageous departure. Acknowledging that the approved and dominant art practice of the time was a detached minimalism, it is little wonder that feminist artists' 'insistence on prioritising experience and meaning over form and style' was considered revolutionary.[26] It was not, however, simply the challenge to dominant form and content that was revolutionary; it was also the *specificity* of the content. For the first time, through drawing on personal experience, an identifiably 'gendered' voice was evident in art.

Located in the context of the 1970s, the various potential politics of these performances of the personal can be summarized:

- The fact that women were creating their own self-representations of 'woman' in place of male representations was implicitly political.
- The fact that women were the subjects of their own art, rather than the objects, was political.
- The fact that women were re-defining and re-inventing 'woman' was political.
- The fact that women were transforming (and thereby controlling and changing) their lived experiences into creative products was political.
- The fact that women were revealing previously hidden or silenced female experiences was political.
- The fact that this revelation simultaneously revealed that the previously considered 'neutral' art was, in fact, equally gendered was political.

- The fact that, through drawing on personal experience, women were forging and practicing new forms of performance was political.
- The fact that these performances (which required only the 'self') were cheap to make, and therefore economically viable, was political.
- The fact that the work may have raised the consciousness of the spectators was political.
- The fact that the work intended to engender community was political.[27]

My intention here is not to map out historical shifts in women's 'autobiographical performance'; 'types' of performance, for good political reason, endure across time, and cannot therefore be identified as 'belonging' to one period.[28] That said, it is undeniably the case that many works from the late-1980s onward are more obviously self-reflexive in their *re*-presentation of the 'I'. Performers such as Lisa Kron, Annie Sprinkle and Bobby Baker explicitly draw attention to the performativity of subjectivity and the construction of the self or multiple selves through and in autobiography. Many performers aim to avoid suggesting any simple relationship between 'a life' and its representation; or even of the 'authority' of experience, understanding experience as always already culturally inscribed, itself an interpretation and in need of interpretation.[29]

Strategies enabling the simultaneous use and critique of 'identity' and 'experience' are as varied as the performances, but have included the deliberate layering of performance modes so that the distinctions between 'truth' and 'fiction' or between 'character' and actor, 'self' and 'other' become blurred and hard to maintain. Narrative trajectories and 'selves' might be multiple and shifting, putting the 'authenticity' of both the performer and the event under scrutiny, with certainties indefinitely postponed. The performer may hesitate, lie or refuse to confess all. Such tactics potentially serve to avoid positioning any performer or narrative as absolute, knowable or essentialized.

Though many performances employ 'destabilizing' strategies, they also nevertheless continue to have, or make, political appeal, walking a fine line between reifying 'real' experience and erasing it. Whilst there may be no stable, essential identity, nevertheless the 'selves' that we variously inhabit continue to feel, variably, the real impact of material conditions. It is such lived materiality that autobiographical performances insist upon and render visible. Performances that draw on the personal, offering particular perspectives, are still often tools of consciousness-raising. Thus Robbie McCauley's *Sally's Rape* (1992) narrates the story of her great-great-grandmother in order to enable a dialogue around race that acknowledges a history of slavery in which we are all implicated and which I/we must speak about and own; Bobby Baker's *Drawing on a Mother's Experience* (1988) makes visible the hidden work of the housewife-mother and the difficulty of being a mother-artist, whilst *Box Story* (2001) recognizes the burden of 'guilt' that women too often unconsciously shoulder; Annie Sprinkle's *Post Porn Modernist* (1989) challenges the good girl/bad girl binary that serves to construct all women as one or the other; Susan Lewis's *Ladies Falling* (1995) inscribes race into the institutions and mannerisms of 'Great Britain', revealing the British 'Lady' as dependent on her 'black other'; Linda Park-Fuller's *A Clean Breast of It* (1993) educates the audience about the experience of breast cancer; Lisa Kron's *2.5 Minute Ride* (1996) recounts her father's story inside the event of the Holocaust but it also tells a story about the difficulty of telling the story, the weight of bearing testimony and of being a reliable witness.[30] It is in the face of such evidence, the work itself, that I have consistently argued that performance of the personal is implicitly political and always necessary.

Is the personal always political?

In October of 2004 I spend some time in New York City attempting to get a 'feel' for the contemporary field of autobiographical performance. I talk to a number of performance artists and programmers of experimental spaces and am perplexed to learn that 'It's not really happening any more', that 'It blossomed with identity politics in the 80s and reached its peak in the 90s', 'that in the 1970s it was a popular mode of performance for economic reasons, but the situation has changed somewhat now and people are turning more to collaborative performance' or that 'It's been seen, heard and done' and 'It's not what's interesting now.'[31] Faced with such responses, it is with an element of surprise that I confront the fact that autobiographical performance, in actual fact, appears to be everywhere. That is, everywhere but the experimental spaces where it first started.

My visit coincides with an annual women's theatre festival, Estrogenius, which includes an afternoon of 'solo' works.[32] A number of these are autobiographical in content. Holly Rizzuto's *Ayai, Yai, Yai* tells her story of being an Italian-Lebanese actress who wants a part in a sitcom (but has the wrong shape of nose), and who ends up (happily ever after) marrying a Jewish man. *Waiting*, a musical by Jenny Mercein, is similarly about an actress trying to make it, relating Mercein's experiences of being an actress who works as a waitress in the Tribeca Grill while waiting both for her big break and for the right man. The publicity for performances programmed as part of the New York Fringe Festival (2004) is also instructive:

> *All the Help You Need: The Adventures of a Hollywood Handyman*: 'A hilarious and frightening autobiographical one-man show. A NY actor hires himself out as a jack-of-all-trades in Hollywood.'
>
> *Becoming Women*: 'Travel back to the South with us as we explore first kisses, first dates, love, our mothers, rape and innocence with humor and intensity. This original two-woman show spot-lights all the bumps and bruises from our years of becoming.'
>
> *Black Martian*: 'Haitian-American, in search of identity, trips over himself encountering psycho cooks, bible-toting street "playas", and knife-wielding Italian women. Journey through 3 countries and 15 characters in this coming-of-age one-man show based on a true story.'
>
> *Confessions of a Mormon Boy*: 'Boy Scout, missionary, NYU-graduate, husband, father... Manhattan escort. After excommunication, divorce, prostitution, and drugs, a sixth-generation Utah Mormon reclaims his "Donny Osmond" smile. A sexy, harrowing autobiographical one-man play told with humor, song, and the Book of Mormon!'
>
> *Daddy Was the Biggest Stage-Mother In Texas!*: 'An autobiographical comedy about a boy who, despite his father's meddling, pursues his dream of becoming a professional dancer. "Tappin' Jack" exists and his heart-warming story is filled with laughter, tears and tap-dancing.'
>
> *Decoding the Tablecloth*: 'Autobiographical solo show examines growing up Jewish and Latina in NY. Koehn plays 20+ characters in this black comedy. From the Warsaw ghetto, to Tango Halls of Argentina, to Disco 70's Brooklyn.'

These are only the entries from A–D. Though I have not seen any of the shows, from the publicity I deduce that (a) there *are* lots of live performances that draw on personal material, (b) that these shows are typically presented as being 'honest', 'humorous' and 'entertaining', and (c) they are now typically performed off-off-Broadway rather than in more experimental performance spaces such as PS122 or Dixon Place.

Reflecting on a number of other off-Broadway performances, which draw explicitly and primarily on personal experience, such as Ellen

Gould's *Bubbe Meises, Bubbe Stories* (1993), Charlayne Woodard's *Pretty Fire* (1993), and Elza Zagreda's *Corn Bread and Feta Cheese: Growing up Fat and Albanian* (2004), I am aware that these performances are formally predictable. Where the use of personal material was once considered radical, these 'solo memoirs' are now almost 'generic', with life events organized in relation to a particular narrative trajectory that unfolds chronologically through time to the present moment. Beginning leads to middle leads to now, where 'now' is represented as the 'happy ending'. In this sense, the model employed bears resemblance to the more traditional model of autobiography, which 'came to be equated with a developmental narrative which orders both time and the personality according to a purpose or goal'.[33] The 'goal' that the performances strive toward is that of everything working out, a personal perspective that typically returns us to the *status quo* via a clichéd 'humanism'. We are not encouraged to think beyond the theatrical experience.

Ellen Gould's *Bubbe Meises* most visibly makes autobiographical performance the stuff of Broadway fare, intercutting each spoken segment with a musical number. The focus of the piece is Gould's inability to make a choice, in a world filled with choices; specifically, choosing to say yes to a marriage proposal. Drawing on the various words of wisdom and stories of her grandmothers', Gould's own final offering of wisdom to the spectator is 'As a human link in a family chain, we will never die.' Charlayne Woodard's *Pretty Fire*, though it figures more directly political events, including the setting alight of a cross by the Ku Klux Klan, dissipates any hint of politics by the end. Woodard's 'memoir', which begins with her premature birth (from which she was not predicted to survive), closes with her 'revealing' why she became a performer. Her grandmother's final wish, in fact made years before she died, was that one of her grandchildren would sing a solo in Church. Woodard obliges, and whilst singing she feels the 'Spirit of the Lord' pass through her. The performance closes with the lines, 'Ever since that Sunday I've been on stages sharing God's gifts and all the while marvelling... Thank you Grandma. Thank you God. Thank you, Ladies and Gentlemen.'

Elza Zagreda's *Growing Up Fat and Albanian* weaves together two main strands – being Albanian in America, and being a woman in Albanian culture. The politics of both are ostensible and explicit; Americans do not know where Albania is – 'Albany? Albino?', and in traditional Albanian culture women have no agency in their lives. And yet, at the end of the performance, we are assured of/reassured by the 'happy ending' as Zagreda (who is, incidentally, blonde and slim) gets married; not to an Albanian, to be sure, but her Albanian family nevertheless accept her choice of husband: 400 of Zagreda's relatives attend the Albanian wedding ceremony:

> And there they were, my family.... Laughing. Happy. Nobody died or got stoned because I married a Latin American.... Marrying an Albanian doesn't make you an Albanian. Albanian means knowing your traditions. Giving your women a hard time and not enough credit.... Albanian is about family, about life, about bonds never being broken, no matter what you do.

Earlier in the performance, Zagreda reveals that she has written an MA thesis on Albanian women's experiences of depression, using their testimonies as her evidence. At the end of the piece she claims to have 'borne witness to a generation of women who didn't have choices.... Women who didn't have a voice. Well, now I hope they do.' It is difficult, however, to 'hear' any dissenting voice, for family and culture remain sovereign, irrespective of whether it is precisely family and culture that are the roots of one's oppression. Throughout the performance a link is made between anger and eating, with fatness becoming the metonym of anger. Yet the performance closes with the lines: 'When all else fails and there's nothing left to do, just go stuff yourself with cornbread and feta cheese. It always made me feel better.' It is difficult to understand how this performance aids any woman, other than the performer, in claiming agency. Zagreda's very specific 'happiness' appears finally to have been located in marriage and family acceptance, she of her Albanian family and their 'quirky' traditions and they of her choice of husband.

The politics of appropriation

Confronted with *Bubbe Meises, Pretty Fire* and *Corn Bread and Feta Cheese,* one cannot help but think that what was initially intended as a radical, challenging and varied practice, in terms of content, form and purpose, has been appropriated and adapted for a different context, with very different aims and outcomes. A number of programmers, when asked about the appearance of autobiographical work on off-Broadway and Broadway, felt that too often actors tended to perceive the solo, autobiographical form as being 'easy' to do, and an opportunity to show their performance skills and talents.[34] According to this view, the autobiographical performance had become a vehicle towards something else, rather than being in itself politically necessary. The 'solo memoir' was simply a calling card. Positioned as a vehicle for performers, a stepping-stone to other work, the form is dictated by future (Broadway) aspirations – hence the frequent use of the traditional narrative model, the inclusion of musical numbers, and the portrayal of a range of 'characters', all of which show off the actor's virtuosity and versatility. It is presumably against such a backdrop that Jonathan Kalb considers solo performances as 'typically performed by frustrated and mediocre New York actors trying to jump-start their me-machines with sitcom-shallow autobiographical monologues'.[35]

This unequivocal reading of more commercially placed autobiographical performances as being less (than) political, might suggest the play of a different sort of politics, a return of a too-simple high – low binary. We would do well to remember that the context for this work is Broadway. Reading Zagreda's performance from a different, more generous angle one could argue that she deliberately and strategically uses a conventional narrative form in order that she might then slip the politics in where they do not typically belong. Though the performance ends on a note of resolution, along the way Zagreda has left us in no doubt of the difficult situation of women in traditional Albanian society, nor of the marginalization and oppression of Albanian people in America (where her father worked in a hospital and 'spent his whole life cleaning up other people's shit'). For Americans, Albanians, American-Albanians and indeed everyone in the audience, the bitter after-taste of various oppressions might just linger in the mouth longer than the final spoonfull of sugar (which is not to deny that this has helped make the political challenges go down).

Seen it, heard it, done it...

Though it would be foolhardy to extrapolate too widely from the experience of New York, it is notable that LA-based performer Denise Uyehara similarly proposes that performers have abandoned the solo, 'autobiographical' form because, since 'co-opted', it is no longer able to serve their purpose.[36] Autobiographical performance in the United Kingdom has never rivalled the extent of the practice in the United States and there is certainly no identifiable trend of the 'solo memoir'. Performances that I have seen recently in Britain retain their political aims. Joey Hateley's *A:Gender* (2003) challenges the gender–sex binary, aiming to carve a place to be something other than 'other', Bobby Baker's most recent work *How to Live* (2004) proposes a form of 'therapy', drawing on her own experiences of mental health and mental health provision. Curious.com's double bill (2004), *Smoking Gun* and *Family Hold Back*, as explored elsewhere in this anthology, work from the micro-political to the macro-political.

The personal mode does also continue to be used in performances in the United States in ways that extend more visibly beyond the resolved 'individual narratives' of the 'solo memoirs' discussed here. Kim Ima's *The Interlude* (2004) recites the history of her father's experience of Japanese internment camps in the United States during World War II, because 'now, more than ever, it is important to remember'. Joni L. Jones *Sista Docta* 'examines the complexities of being an African American woman professor in predominantly European American institutions'.[37] Though first performed in 1994, Jones has continued to perform this piece over the past ten years. Linda Park-Fuller also continues to perform *A Clean Breast of It* whenever she is invited. These performances suggest that the telling of certain stories remains necessary, for different people, at different times and in different places. Where some people may purport to 'have heard it, seen it, done it', others might be desperate to hear and see it, and some women may need to perform it. What is considered political in one place may seem irrelevant in another, putting a slightly different spin on Harris's statement that 'not all of the personal is political in *exactly the same way* and to the same effect.'

An unavoidable context for all contemporary performances that draw on personal material is the wider contemporary Western culture, saturated as it is with 'confessional' opportunities provided by 'reality television' shows too many to mention. If Foucault considered us 'confessing animals' in 1976, by the twenty-first century we might justly be accused of having become 'confessing monsters'.[38] 'The personal is political' might mean something different against such widespread commodification of the personal, where inciting 'personal revelation' becomes a strategy to ensure ratings success in a competitive viewing market. Though there is no space to discuss it here, this exploitation of the personal prompts its own political consideration.

The sheer availability of mass-mediated personal narratives, what Jon Dovey refers to as 'a popular ethnography of the everyday',[39] has arguably rendered the autobiographical performance an ineffective political tool. In an environment where subjective experiences are now the 'new regime of truth' (*FREAKSHOW*, p.25), the autobiographical performance is simply one more first-person narrative in an indistinguishable cacophony. Against this 'mass', challenges to 'form' become primary. The incorporation of the personal into performance was, from the outset, formally challenging, with formal experimentation grounded in the political aims of the work, framed by the questions 'what do you want to say?' and 'what is the best way to do that?' In an era in which the personal has become inseparable from the public (aided also by the issue of identity cards, retina and fingerprint scanning, CCTV, etc.), in order that we are enabled to see it any performance of autobiography would do well to differentiate itself. As one programmer observed, few people seem interested any longer in 'pushing the envelope'.

However, formal challenges do persist. Curious.com and Lois Weaver's *On the Scent* (2004), for example, transplants their personal experiences by performing them in other people's bedrooms, living rooms and kitchens. Their personal stories, located in others' personal spaces, generate a surprising and unscripted dialogue. Another recent 'push' to the form is the move away from solo performance. In Lisa Kron's *Well* there are five other performers, one of whom plays her mother. Notably, the actors also all play 'actors'. Kron's aim, of destabilizing the 'omniscient narrator', is enabled by these 'characters'/actors who continuously throw her play off-course. *Well*, as a result, is a meta-theatrical, self-reflexive *tour de force*. In the opening moments of the play, 'Kron' self-consciously informs us that

> This play is *not* about my mother and me.... It's also not about how she's been sick for years and years and years and I was sick as well but somehow I got better. It's not about how she was able to heal a neighbourhood but she's not able to heal herself. It's not ABOUT those things but it does use those things as a vehicle for a *(reads off a note card)* 'theatrical exploration of issues of health and illness both in the individual and in a community'.

Rather than assuming that an 'autobiographical form' is simply available to be appropriated, Kron's challenge, since she began making work, has been to address the fact that 'first person solo work isn't inherently suited to theatre'. Simultaneous with using personal experiences to bear witness to wider sociopolitical events, her work is 'about solving those formal problems'.[40]

Kim Ima's *The Interlude* (2004) similarly departs from the solo form, using an additional six performers to restage the events (historical and imagined) of her father's experiences in the internment camp. This turn to the more evidently theatrical (the 'play') is perhaps one further response to the recognition of the construction of the 'I', of the 'writing'/performing of a 'self' and a 'life'. Though certain 'authenticating' tools remain, such as black and white photographs of her father as a young boy, the reconstruction of the story, through the involvement of a cast, becomes more evident. Given that Ima's father rarely spoke about the experience (and perhaps had only vague recollections), a visible recuperation becomes a conscious and deliberate *act* of remembering (in the face of refusals to even acknowledge). Both *Well* and *The Interlude*, moving away from solo performance, might be thought of as pushing the form of autobiographical performance. Though my focus here is on work made by women, it is also notable that a number of mixed-gender collaborative companies have presented collective 'auto/biographies', works which draw on their own (presumed) personal experiences and/ or solicit others'. Forced Entertainment's *The Travels* (2002) and The Civilians *Nobody's Lunch* (2004) are just two recent examples.

Keeping it real

Whilst the past ten years have witnessed vigorous debate concerning the existence, and difference, of a 'Third-Wave' feminist movement, one belief that appears to have endured across perceived shifts (assumed, real or contested) is the implicit value of personal experience. Given the thorough challenge to notions of 'authority' and the 'authenticity' of experience enabled by poststructuralist feminism, it is tempting to read this as a 'return of the real'. The title of Rebecca Walker's 1995 anthology, *To be Real: Telling the Truth and Changing the Face of Feminism*, when held against the critical backdrop in which experience is unable to easily signal beyond the frame of its own text, comes as something of a surprise.[41] Walker's introduction, 'Being Real', draws on her personal experiences as a means to make sense of her relationship to feminism. This use of the 'personal' is a trend sustained by the other contributors; being real, it would seem, is to tell it like it (really) is: 'The writers here have done the difficult work of being real (refusing to be bound by a feminist ideal not of their own making)' (Walker, *To be Real*, p.xxxiv).[42] Implicitly connecting with 'Second-Wave' praxis, Walker attests that 'our lives are the best basis for feminist theory...' (p.xxxvii). An unproblematic relationship between ontology and epistemology is similarly reinstated in Barbara Findlen's *Listen Up: Voices from the Next Feminist Generation*, published in the same year as Walker's collection. Placing personal experience as 'foundational', Findlen explains that 'Individual women's experiences of sexism have always been an important basis for political awareness and action. This collection gives voice to young feminists' personal experiences because they have often been, and continue to be, our point of entry into feminism.'[43] Leslie Heywood and Jennifer Drake, meanwhile, in *Third Wave Agenda: Being Feminist, Doing Feminism*, pay homage to Gloria Anzaldúa and Cherie Moraga by insisting on '"theory in the flesh"' and '"lived theory"', which can 'articulate the historically situated experiences of our "generation"', 'good storytelling *and* critical analysis without jargon' (Heywood and Drake, p.14). A more recent publication, *Manifesta: Young Women, Feminism, and the Future,* similarly theorizes from the personal; the material of the first chapter is generated from a dinner party held by the authors, which 'is just as likely to be a place to see politics at work as is a rally . . . because, as with every wave of feminism, our politics emerge from our everyday lives' (Baumgardner and Richards, p.15).

The inheritance of a feminist praxis rooted in personal experience and consciousness-raising activity is explicitly recognized by these younger feminist writers. Explaining their opening 'dinner party' chapter, the authors of *Manifesta* note that the concept 'came out of a fierce faith that this honest communicating among women is a revolutionary act, and the best preface to activism' (Baumgardner and Richards, p.15).[44] Rory Dicker and Alison Piepmeier, editors of *Catching a Wave: Reclaiming Feminism for the 21st Century,* similarly claim the second-wave strategy of consciousness-raising, framing their own collection as a consciousness-raising tool: 'Because backlash rhetoric and our complacency have inured many of us to inequalities that persist around us, we wanted to show our readers how to wake up to these injustices and begin to do something to redress them' (Dicker and Piepmeier, pp.4–5). Baumgardner and Richards explicitly identify the 'lack of consciousness [as] one reason that the [feminist] movement is stalled' (*Manifesta*, p.83) and insist that:

> Testimony is where feminism starts. Historically, women's personal stories have been the evidence of where the movement needs to go politically and, furthermore, that there is a need to move forward.
> (*Manifesta*, p.20)

All of these examples testify to Whelehan's recognition that consciousness-raising is a component of feminist pedagogy, and 'can itself be rejuvenating for feminism' (*Modern Feminist Thought*, p.198). Where there is an 'assumption that feminist politics are redundant, then "consciousness raising" is again one of the most vital feminist activities' (p.241).[45]

It is perhaps more than coincidence that this return to the personal as a mode of consciousness-raising, in spite of its commodification in mass-media forms, is simultaneous with the wider political shift towards a radical conservative agenda. As 'personal' rights become increasingly threatened or denied (a woman's right to choose, the right to be critical of government decisions, the right to practice affirmative action, the right to have one's lesbian relationship legally recognized), the politics of the personal become more easily identifiable. The personal might be considered most political in those moments and places where it is most under attack. In such places and times, attention to and upon the personal, for the purposes of consciousness-raising and the performance of counter-narratives, remains a strategic response. Of course, this perceived 'return' to the 'personal' might simply be another example of the commodification of personal experience. Given the primacy of context, it is impossible to definitively proclaim the politics of any use of the personal. Whether a performance 'matters', then, and in what way, depends not only on the matter of its content, but also on who makes it, who witnesses it, where, and when.

Notes

1 The research for this chapter has been supported by the AHRC, to whom I am grateful. My warmest and sincerest thanks also to the numerous people who shared their thoughts and archives with me.
2 Holly Hughes, in Holly Hughes and David Román, eds, *O Solo Homo: The New Queer Performance* (New York: Grove Press, 1998), p.8.
3 The term 'autobiographical performance' is slightly misleading. Though feminist performances that draw on personal experience do share political and strategic ground with written feminist autobiographies, they also differ in that performances are not often concerned with attempting to present a life story or life history (or even deliberately challenging such a practice). Their 'autobiographical' scope is usually narrower.
4 Geraldine Harris, *Staging Femininities: Performance and Performativity* (Manchester: Manchester University Press, 1999), p.167.
5 Leslie Heywood and Jennifer Drake, eds, *Third Wave Agenda: Being Feminist, Doing Feminism* (Minneapolis: University of Minnesota Press, 1997), p.23.
6 Katha Politt and Jennifer Baumgardner, 'Afterword: A Correspondence between Katha Politt and Jennifer Baumgardner', in Rory Dicker and Alison Piepmeier, eds, *Catching a Wave: Reclaiming Feminism for the 21st Century* (Boston, MA: North Eastern University Press, 2003), pp.309–19, pp.317–19.
7 Imelda Whelehan, *Modern Feminist Thought: From the Second Wave to 'Post-Feminism'* (Edinburgh: Edinburgh University Press, 1995), p.13.
8 Robin Morgan, cited in Jennifer Baumgardner and Amy Richards, *Manifesta: Young Women, Feminism, and the Future* (New York: Farrar, Strauss & Giroux, 2000), p.19.
9 See Sheila Rowbotham, Lynne Segal and Hilary Wainwright, eds, *Beyond the Fragments: Feminism and the Making of Socialism* (London: The Merlin Press, 1979); Sheila Rowbotham, *The Past is Before Us: Feminism in Action Since the 1960s* (London: Pandora Press, 1989).
10 Bonnie Zimmerman, 'The Politics of Transliteration: Lesbian Personal Narratives', in Estelle B. Freedman, Barbara C. Gap, Susan L. Johnson,

Kathleen M. Weston, eds, *The Lesbian Issue – Essays from Signs* (University of Chicago Press: Chicago,1985), pp.251–70, pp.253–4.

11 I am using these 'periodic' designations because they have already become shorthand reference, although I recognize that the concept of 'waves' is problematic. See Siegel; Whelehan; and Stacy Gillis, Gillian Howie and Rebecca Munford, eds, *Third Wave Feminism: A Critical Exploration* (Basingstoke: Palgrave, 2004).

12 Kristina Sheryl Wong, 'Pranks and Fake Porn: Doing Feminism My Way', in *Catching a Wave*, pp.294–307, p.296.

13 Irene Gammel, ed., *Confessional Politics: Women's Sexual Self-Representations in Life Writing and Popular Media* (Carbondale and Edwardsville: Southern Illinois University Press, 1999).

14 Sidonie Smith, 'The Autobiographical Manifesto: Identities, Temporalities, Politics', in Shirley Neuman, ed., *Autobiography and Questions of Gender* (London: Frank Cass, 1991), pp.186–212, p.189.

15 Sidonie Smith, 'Autobiographical Manifestos', in Sidonie Smith and Julia Watson, eds, *Women, Autobiography, Theory: A Reader* (Wisconsin: University of Wisconsin Press, 1998), pp.433–40, p.434. Originally published in Sidonie Smith, *Subjectivity, Identity and the Body* (Bloomington and Indianapolis: Indiana University Press, 1993).

16 bell hooks, *Talking Back: Thinking Feminist, Thinking Black* (Boston, MA: South End Press, 1989), p.43.

17 Liz Stanley, *The Auto/Biographical I: The Theory and Practice of Feminist Auto/Biography* (Manchester: Manchester University Press, 1992), p.116.

18 Kenneth Plummer, *Telling Sexual Stories: Power, Change and Social Worlds* (London and New York: Routledge, 1994).

19 Bobby Baker, Interview with the author, London, 2001.

20 Lenora Champagne, 'Notes on autobiography and performance', *Women & Performance: A Journal of Feminist Theory*, 19: 10 (1999): 155–72, p.155.

21 Elizabeth Bell, 'Orchids in the Arctic: Women's Autobiographical Performances as Mentoring', in Lynn C. Miller, Jacqueline Taylor and M. Heather Carver, eds, *Voices Made Flesh: Performing Women's Autobiography* (Wisconsin: University of Wisconsin Press, 2003), pp.301–18, p.315.

22 Jeannie Forte, 'Women's performance art: Feminism and postmodernism', *Theatre Journal*, 40 (1988): 217–35, p.224.

23 Linda Park-Fuller, 'A Clean Breast of It', in *Voices Made Flesh*, pp.215–36, p.215.

24 Tami Spry, 'Illustrated Woman: Autoperformance in "Skins: A Daughter's (Re)Construction of Cancer' and 'Tattoo Stories: A Postscript to Skins"', in *Voices Made Flesh*, pp.157–91, p.169.

25 Faith Wilding, 'The Feminist Art Programmes at Fresno and Calarts, 1970–75', in Norma Broude, and Mary D. Garrard, eds, *The Power of Feminist Art: The American Movement of the 1970s, History and Impact* (New York: Harry N. Abrams, 1994), pp.32–47, p.34.

26 Broude and Garrard, *The Power of Feminist Art*, p.10.

27 See Broude and Garrard, *The Power of Feminist Art*; Helena Reckett, ed., *Art and Feminism* (London: Phaidon Press, 2001); Terry Wolverton, *Insurgent Muse: Life and Art at the Woman's Building* (San Francisco: City Lights Books, 2002); Jeannie Forte, 'Women's Performance Art'; Moira Roth, 'Autobiography,

Theater, Mysticism and Politics: Women's Performance Art in Southern California', in Carl Loeffler, ed., *Performance Anthology: Source Book of California Performance Art* (San Francisco: Contemporary Arts Press, 1980), pp.463–89; Judith Barry, 'Women, Representation, and Performance Art: Northern California', in *Performance Anthology*, pp.439–62; Catherine Elwes, 'Floating Femininity: A Look at Performance Art by Women', in Sarah Kent and Jacqueline Morreau, eds, *Women's Images of Men* (London: Writers and Readers Publishing, 1985), pp.164–93.

28 Retrospective readings of this work sometimes reposition it as dangerously essentialist and universalizing, and naive in relation to concepts of 'identity' and 'experience'. However, the past ten years have witnessed something of a reappropriation of the political credibility of this early feminist work. See the various contributions in Broude and Garrard, *The Power of Feminist Art*, and Reckett, *Art and Feminism*.

29 See Joan W. Scott, 'Experience', in Judith Butler and Joan W. Scott, eds, *Feminists Theorize the Political* (London and New York: Routledge, 1992), pp.22–41.

30 Each of these examples obviously does much more than I give them credit for here. Some anthologies of 'performance scripts' have recently been published. See Hughes and Román, eds, *O Solo Homo* (1998); Jo Bonney, ed., *Extreme Exposure: An Anthology of Solo Performance Texts from the Twentieth Century* (New York: Theatre Communications Group, 2000); Alina Troyano, *I, Carmelita Tropicana: Performing Between Cultures* (Boston, MA: Beacon Press, 2000); Mark Russell, ed., *Out of Character: Rants, Raves, and Monologues from Today's Top Performance Artists* (London: Bantam Books, 1997); Miller, Taylor and Carver, *Voices Made Flesh*; Moira Roth ed., *Rachel Rosenthal* (Baltimore, MD: The John Hopkins University Press, 1997); Sydney Mahone, ed., *Moon Marked and Touched By Sun: Plays By African American Women* (New York: Theatre Communications Group, 1994); Lenora Champagne, ed., *Out from Under: Texts by Women Performance Artists* (New York: Theatre Communications Group, 1990).

31 Some interviewees wished their comments to remain anonymous and for this reason I have left these statements uncredited.

32 Manhattan Theatre Resource, 23 October 2004.

33 Linda Anderson, *Autobiography* (London and New York: Routledge, 2001), p.8.

34 In reality, of course, the solo performer who creates their own performance based on personal material must be both a good script-writer *and* performer.

35 Jonathan Kalb, *Theater*, 31: 3 (2000), p.14.

36 Email communication, 16 November 2004.

37 Joni L. Jones, 'Sista Docta', in *Voices Made Flesh*, pp.237–57, p.237.

38 Michel Foucault, *The History of Sexuality, Volume 1: An Introduction*, trans. Robert Hurley (London: Penguin Books, 1990 (1976)), p.59

39 Jon Dovey, *FREAKSHOW: First Person Media and Factual Television* (London: Pluto Press, 2000), p.138.

40 Interview with the author, 2004, New York.

41 Rebecca Walker, ed., *To Be Real: Telling the Truth and Changing the Face of Feminism* (New York: Anchor Books, 1995).

42 This bears interesting relation to 'second-wave' feminist attempts to replace what was perceived as 'men's' vision of 'women' with more 'real' representations.

43 Barbara Findlen, *Listen Up: Voices from the Next Feminist Generation*, 2nd edn (Seattle: Seal, 2001), p.xv.
44 Baumgardner and Richards also signify their feminist heritage by making direct reference to Judy Chicago's influential art installation 'The Dinner Party' (1974–9).
45 In her project with female students, *Self-ish* (1997), Elaine Aston uses autobiographical performance-making as part of a feminist pedagogical practice. See Elaine Aston, 'Staging Our Selves', in Alison Donnell and Pauline Polkey, eds, *Representing Lives: Women and Auto/Biography* (Basingstoke: Palgrave Macmillan, 2000), pp.119–28.

10

Performing in Glass: Reproduction, Technology, Performance and the Bio-Spectacular

Anna Furse

Ex ovo omnia (everything from an egg)
(Harvey, *De Generatione Animalium, 1654*)

'but I'm a woman thinking egg
and staggering under its weight'[1]

Walking on eggs

The human egg is a human future. More precisely, the human egg is a potential human future. A baby girl carries all the eggs she will ever need in a lifetime inside her ovaries.[2] And inside each of these eggs is the genetic material which, if fertilized, could make another baby girl, who contains the potential future of another baby (girl) inside her *ad infinitum.* A woman's future in her own lifetime always contains the issue of whether or not she will use her eggs to become a mother. What this might mean is the matter of human drama, both on stage and off. In this chapter I will try and locate some of these dramas as they have been written for the stage, from a contemporary (feminist) perspective and with specific reference to the spectacular in In Vitro Fertilization (IVF)[3] and Assisted Reproduction Technology (ART).

In the feminist future reproductive issues will remain high on the agenda as this will doubtlessly shift increasingly from the 'natural' to the technological. ART is joining key feminist autonomy and choice themes – contraception and abortion – that each continue to signify that a woman's body is the site on which many political battles are faught. Control over women's bodies is a human rights issue worldwide as the

Egyptian doctor, writer and militant Nawal El Saadawi reminded her audience in a talk at Toynbee Hall on International Women's Day, 2005.[4]

One thing is certain when discussing reproduction: the subject unites all of humanity. Whether it is having as many babies as possible as survival strategies in poor countries where infant mortality rates are high, or the issues of 'choice' for those living in the richer Northern hemispheres, every person, whatever their creed, sexuality or environment, develops an attitude to having children or not. Choice may not be ours at all. Nature may decide for us. A pregnancy that is desirable or has been actively tried for might end in miscarriage, even so early on that they are sometimes mistaken for periods.

Reproductive science tries not so much to cheat Nature as to set, opportunistically, the optimum conditions for a favourable result. In the globally expanding industry of ART and reprogenetics, experts manipulate hormones, gametes, embryos, timing, and even the relationship between genetic and birth mothers in varying combinations in an attempt to provoke women's bodies into successfully releasing and receiving eggs and nourishing embryos. In this new technological Eden[5] the human egg is becoming a site for medical and technological intervention. The egg, like other single cells, can now be isolated, manipulated, traded and visualized. It can be chemically triggered to release itself into the ovary in multiples (super-ovulation), injected with a single sperm (ICSI),[6] stripped of its nucleus and be injected with donor cells (cloning), moved from one woman to another (egg donation) or commodified (trade in gametes). Finally, the infintessimally small human egg can be pictorialized via ultrasound sonogram or micro-digital technology.

In the richer democratic countries the choice of not reproducing, of being child*free* (not child*less*) compound economic, emotional, sexual and lifestyle issues. If the choice is motherhood, ART offers new options for the sub-fertile. This relatively recent branch of medicine is controversial for feminists. It might signal a patriarchal pronatalist strategy for maintaining a *status quo*, a male-dominated anti-natural medical intervention in the woman's body, a privilege for the rich and those deemed reproductively 'fit', or a tool for choice with significant potential for social change.[7] Zoologist Robin Baker imagines futuristic scenarios and 'reproductive restaurants' where choices on the baby-making menu demonstrate the ways in which ART dislodges so many motivations towards conventional heterosexual coupling. Baker argues that with accelerating developments in ART and reprogenetics, whilst men become unnecessary for their sperm and women might not even need their own eggs and wombs, creative permutations in future family construction would become commonplace. He predicts that ART, far from reducing our sex lives to anodyne arrangements between consenting parties, will put the swinging 1960s back into our lives, but free from VDs and the risks of pregnancy. Sex, finally irrevocably divorced from reproduction, will be simply sheer fun as 'reproductive technology will unleash the ancient urges behind human sexuality to a degree that has never before been possible'.[8] The reproducer, in this future, is consumer.

Biologist Lee Silver takes up the same theme arguing that the development of reprogenetics is inevitable 'for better *and* worse' and that 'a new age is upon us – an age in which we as humans will gain the ability *to change the nature of our species'* (*Remaking Eden*, p.13). Silver's caution emphasizes just how fraught with ethical debates this new reproductive landscape lies. At the heart of bio-ethical issues lies the fundamental question 'what is human Life and when does it begin?' This tends to be followed by heated debate on what is 'natural'. Such ideas strike deep into personal and collective ideology, linked as they rapidly become to concepts of destiny and creation. Just as with certain pornography debates in the 1980s, fundamentalist feminists can find themselves in odd alliance with the Christian Right. Dion Farquhar usefully reminds us of the important third position in the pro- or anti-ART stance and argues that 'there is a more ambivalent agnostic postmodern position. Instead of demonizing technology, vilifying consumption, and idealizing an edenic natural, it looks at the demographics of ART users – as well as those who cannot even qualify as users because of poverty or cultural isolation from technological medicine, etc. – and suggests *transgressive social and political possibilities*' (my emphasis, Farquhar, *The Gendered Cyborg*, p.216.)

Farquhar introduces the term Other Mothers to describe all those offered the opportunity to parent by ART – single women, gay men and lesbian couples. She argues forcefully that rather than deploring the demise of the nuclear family, the interference with Nature and the disturbance to the traditional norm of what age and type a mother should be, ART might be celebrated instead as contributing to the 'proliferation of Other kinds of families and the growing fissures in the near-hegemonic figure of the mother' (*The Gendered Cyborg*, p.217). The point remains that proper and fair access to such treatments for all, and within a National Health Service, is the only way forward if such major social changes are to impact across societies as a whole and not just privilege the socially advantaged.

Whilst such crucial discourses are conducted at the political theoretical level, ordinary women with ordinary lives, drives, urges or pressures towards motherhood when facing their own sub-fertility and involuntary childlessness may, if given the opportunity, 'choose' in despair to gamble with the uncertain outcome of their ART.

Wide-awake conception

IVF blends the most simple and the most complex ideas we could possibly have about the (reproductive) body. On the one hand, the process of extracting eggs and sperm, introducing them to each other in laboratory conditions and putting the fertilized embryo(s) back in the woman's womb is as vividly simplistic as a child's drawing. After all, how are babies made? Gametes have to meet, that's all. On the other hand, complex ideas cluster here. The technology is 'high', involving micro-electronics 'incorporated' into the process. It employs extraordinary imaging technology and offers new insight into the body. We are thrust into a provocatively new relationship with ourselves, our sexual bonding, our ideas about life and its origins as the hitherto invisible miracle of life presents itself spectacularly before our eyes, with imaging tools now offering not a generalized, still picture of our inner workings as in past technologies such as X-Ray, but portraits of the body as individual, evolving, mobile.[9] With IVF both the visible and the imagined encourage pragmatism. The romance, pleasure or accident of a spontaneous conception 'made in bed' is replaced by a highly programmed, meticulously scheduled procedure. You think reproduction at its lowest common denominator: single egg and sperm cells. The possibility of separating these cells from their bodily source can lead to their commodification and '[t]he organic unity of fetus and mother can no longer be assumed, and all these newly fragmented parts can now be subjected to market forces, ordered, produced, bought and sold.'[10]

Unlike North America, in the United Kingdom the HFEA[11] forbids trading in sperm, eggs and embryos. Nonetheless the language surrounding the movement of our reproductive material is bluntly materialistic. In the quest of producing what clinics here call 'take-home babies', ART users can produce 'spares', they can 'freeze', 'host' or 'donate'. Reproductive bits have a potential all of their own, even without the individual woman, as private parts become moveable parts. ART users paradoxically dis-integrate the organic self, if temporarily, in a quest to experience the integrity of pregnancy and the ultimate goal, parenthood.

ART is medically highly invasive and nobody enters such a venture lightly. As an ART user,[12] I continue to wonder at how fervently we involuntarily sub-fertiles apply ourselves to the project of reproducing even when faced with risk of failure (IVF has a 25 per cent success rate at best), not to mention expense. The drive may be experienced as biological or psychological or both. Social pressure might have become internalized, for the non-mother is non-normative, she is Other. She is not using her eggs, her potential to contribute to the future. Non-reproductive woman is threatening, destructive, negative, imploding. Sub-fertiles are child*less*. We *lose*. Our eggs are too *old*. We are in a *decline*. We *fail* to conceive. In the patriarchal language of classic medical textbooks menstruation was depicted as 'the uterus crying for the lack of a baby' (Martin, *The Woman in the Body* p.45). The language of menstruation persists as 'loss', 'shedding', 'haemorrhaging' and 'failure'. Martin continues

> [I]t also carries the idea of production gone awry, making products of no use, not to specification, unsaleable, wasted, scrap ... Perhaps one reason the negative image of failed production is attached to menstruation is precisely that women are in some sinister sense out of control when they menstruate. They are not reproducing, not continuing the species, not preparing to stay at home with the baby, not providing a safe, warm, womb to nurture a man's sperm.
>
> (*The Woman in the Body*, pp.46–7)

Deviance

It is vital to distinguish between the childfree and the involuntarily childless woman. The former has to cope with social stigma, but the latter is dealing with *both* this *and* a profound sense of powerlessness and failure as the biological clock pounds. The idea of waste becomes acute. What might have been 'sinister' and out of control in menstruation is now carried over paradoxically into the inability to conceive. In the collective psyche infertiles inhabit a hinterland of difference and non-belonging to the milky, messy, fleshiness of motherhood. Female archetypes of outcasts lurk around the condition, those terrifying non-mothers: witches, hags, hysterics, whores, old maids, evil stepmothers, butches. We (I will deliberately use this pronoun henceforth for I am sub-fertile) are dry, withered, bitter crones. We are simply not *feminine*, not *natural*. We are a waste of eggs, of breast, of womb.

Barren bodies are inscribed with meanings. Barren women recur as deviants throughout Judeo-Christian cultural history as Bible stories tell. Centrally, we are offered here an impossible legacy, a punitive yardstick by which to measure ourselves morally and maternally: the Virgin-Whore paradigm embodied in Mary (divine Donor Insemination?) versus the childless prostitute Magdalene.[13] From Magdalene to Rachel, desperate for a child 'Give me a child lest I die' (Genesis 39:6)[14] the childless woman figures as either outcast, murderous or suicidal. Her pariah status makes her appear in literature as not only desperate but destructive – whether her condition is desirable to her or not. Is the particular threat of the barren, non-reproductive woman an enduring phobia because, as Martin indicates above, she has stepped out of the frame, she is free, she is mobile, she is dangerous, she has the opportunity to experience her sexuality as distinct from reproduction, she is, in short, *manly*?

Pre-ART dramas

Even such proto-feminist or woman-identifying playwrights as Ibsen and Lorca endorse a negative construct of the non-mother woman in the tremendous murderous/suicidal passions they portray in the eponymous desperados Hedda Gabler and Yerma. These characters' non-reproductivity is mirrored in their destructivity. In Hedda, Ibsen give us a complex, intelligent, thwarted woman suffocating in the middle-class ménage into which she has compromisingly married. Whilst her bookish, earnest husband Tesman pores over the history of animal husbandry in medieval Brabant (*sic*), oblivious to his young wife's hysterical tendencies, she drives herself crazy with pent up frustration. Ibsen remains ambiguous as to whether she is in actual fact pregnant following the long honeymoon or whether she is simply terrified of becoming so. In the opening scene of the play Tesman asks his clucking Aunt Julia euphemistically if she has noticed how his new wife has put on weight during their long honeymoon to which Hedda snaps a bitter denial. Hedda's body is discussed as if she were a chattel, a cow being fattened. At any rate, Hedda herself refuses to acknowledge or embody a pregnancy. Thereafter, any latent maternal instincts get subsumed in her jealous rage at the literary child her rival Thea Elvsted has created with the Byronic Lovborg. Launching herself and Lovborg headlong into disaster, she gives him a pistol to shoot himself with and, burning his manuscript in her husband's stove declares 'Your child and Eilert Lovborg's child . . . I'm burning it, I'm burning your child.'[15] Finally Hedda, a self-avowed coward, terrified of scandal and appalled at the mess she has made of things, shoots herself.

Hedda's non-maternality is linked to everything corrosive and explosive. She *is* her father's pistols, phallic and lethal. In his feminist heroine who rejects the constraints of bourgeois morality Ibsen offers us not only a non-mother but a non-possible woman. Hedda's non-conformity cannot survive, and the woman who cannot fulfil herself cannot give birth.

Instead of using her eggs and womb procreatively, she destructively ejaculates the phallus of her father's pistols in bloody suicide.

The Spanish writer Federico García Lorca's eponymous protagonist Yerma continues the theme of the destructive force of a woman with eggs-gone-to-waste. Contrary to the urban, aristocratic Hedda, Yerma is a poor illiterate peasant girl, married but without issue. Yerma, like Hedda, is also desperate, not to get away from it all but to enter fully into her pronatalist community and breed. The word Yerma means 'barren' in Spanish. Projecting himself into the psyche of this childless young woman, Lorca articulates the feeling of being on the outside of a fertile world. The play's poetry is steeped in infertile/fertile imagery – arid landscapes, thistles, heat, in contrast to orchards, water, fountains. Yerma identifies herself with her environment. She *is* the barren landscape in which she lives. She is a parched earth longing for saturation, likened to a bunch of thistles. She is taunted by the fecundity of others, animal, human and vegetable. She cannot distinguish between herself and Nature, and experiences herself as an aberration of its bounty. She is Nature at its worst, useless, impossible to cultivate, without purpose.

As with Hedda another, more sexually lustruous man Victor, appears in the shadows of Yerma's marriage to Juan. Their trysts remain unconsummated. Their tense meeting in a parched field is interrupted by Yerma's womby hallucinations:

Yerma: Listen!
Victor: What?
Yerma: Can't you hear crying?
Victor: (*listening*) No.[16]

Feelings run high in this agenda. Victor offers the promise of full and complete (reproductive?) sex, whereas Juan is cold and unfeeling. Victor even suggests that Juan was always 'dried up' (Lorca, *Yerma*, Act 1, scene 2, p.173) inferring that here the infertility factor is male.

Juan is certainly unconcerned with the lack of progeny: 'Life is simpler without kids, and I like it that way' (Lorca, *Yerma*, Act 3 scene 2, p.205). His lack of identification with Yerma's procreative desires is callous 'this thing means nothing to me' (Lorca, *Yerma*, Act 3, scene 2, p.204) and perhaps defensive. At the end of the play, begging Yerma to stop her 'crying for the moon' (Lorca, *Yerma*, p.204) he floors her in a fight and his desire surges. For Yerma, sex without reproductive purpose appalls her. Juan-husband is also Juan-potential-father-of-a-son, and even a child substitute. It is typical of Lorca to conflate husband and child in ways that resonate meaning. Now, faced with a husband who will not (cannot?) provide her with a baby, thwarted, crushed and enraged by his sweaty desire, she leaps at him and strangles him to death, declaring herself finally 'Barren and alone forever'. Then, yelling to the crowd that gathers round, 'yes I've killed him. I've killed my son' (Lorca, *Yerma*, Act 3, scene 2, p.206).

Lorca writes brilliantly about the utter despair to which the involuntary childless are driven. But the maternally thwarted Yerma, like Hedda the non-maternal, takes it out on others with tragic consequences. The non-mother cannot survive the imperatives of her culture. The eggs which do not fertilize into the flesh of the future lie instead like time bombs exploding from inside these despairing women.

Post-ART dramas

Infertility dramas lie few and far between. A handful of women playwrights have taken up the theme of biological clocks and reproductive options in the face of social/sexual circumstances (being single, gay or lesbian) and available technologies both primitive and advanced. A 1978 play by Michelene Wandor, *AID Thy Neighbour,*[17] dealt with 'turkey-baster' artificial insemination (AI – now clinically regulated as DI, Donor Insemination) for a lesbian couple – in retrospect a drama ahead of its time, dealing as it does not with infertility as a biological impairment, but in reproductive strategies for Other Mothers. Another prescient Wandor play, *Wanted,* is about surrogacy.[18] Liz Lochhead's recent play *Perfect Days*[19] revolves around the reproductive crisis for her protagonist Barbs, a successful 39-year-old professional. Single and childless, defying all conventions, she also uses the turkey baster (she calls 'the middle man') with her gay friend, takes a younger, adopted man as her lover and, eventually, after many relationship complications, has a baby with – as it turns out misidentified – donor sperm.

Lochhead's device of Barb's birthday as a catalyst for the biological-clock drama is also employed by Timberlake Wertenbaker in her 1995 play *The Break of Day*. Wertenbaker, via two simultaneous and interwoven narratives, deals with themes of adoption of Eastern European babies alongside IVF, presenting a prismatic perspective on contemporary issues for those-whose-eggs-have-run-out. Fertiles often accuse ART users of being selfish given all-the-unwanted-babies-in-the-world and *The Break of Day* shows how fraught the adoption process can be. Whilst the character Nina is struggling in an unspecified Eastern European city with critical baby health issues, bureaucracy and corruption, Tess is in an Assisted

Conception Unit. The play questions class, money and what can become of once independent-career-woman-in-the-grip-of-baby-lust. Towards the end of the play Tess, consumed with her quest, is opting for egg donation. Her husband Robert deplores her psychological and professional demise (it is hard for anyone who hasn't been through it to understand how this can be) and begs her to go back to work, 'Partly to pay the mortgage and partly to come home and tell me what you did all day.' To which Tess snaps, 'I can tell you what I do every single day. I walk down the street and leer at women. Which one has the good eggs? . . . Women used to be my sisters. They're objects: egg vessels. Now you know.'[20]

Male partners' inability to completely comprehend the lengths to which a woman's infecundity will take her leads Yerma to strangle Juan and Tess to leave her husband. Infertility is a major life crisis and the strain it puts on relationships intolerable. Ben Elton, with smug political correctness, highlights this stress factor from the male point of view. His film *Maybe Baby*, from his novel *Inconceivable*,[21] attempts a lot of laughs when Lucy and Sam – yuppie media types – embark on Lucy's baby quest. Elton's gallows humour comes from experience, and he certainly feels for women, but he predictably emphasizes that it is she, not he, who loses her mind when driven by her biological clock. If Yerma kills, then the Tesses, Ninas, Barbs and Lucys of the world are prepared to jeopardize career, relationship, lifestyle, even their own health to have a baby of their own. Meanwhile, according to Elton, new men dance indulgently around their possessed womenfolk, donating sperm, tolerating hormonally induced mood swings, riding the roller-coaster, for they know, *willy nilly*, that if they are to keep this woman, they must walk on eggs.

Knowing in our flesh and bones

In 1977 the poet Adrienne Rich announced that feminism offered women the possibility, for the first time in history, the opportunity to convert our embodied experience into both 'knowledge and power'.[22] Philosopher Maurice Merleau-Ponty's argument for the integration of the Cartesian mind/body split into a 'primacy of perception' has put subjectivity into a framework in which the body itself, its function and its senses are foregrounded as a way of understanding the world. For Merleau-Ponty the body is 'no longer conceived as an object of the world, but as our means of communicating with it'.[23] He has been criticized by some feminists for universalizing experience, yet his 'defiance and challenge to binary polarizations places his interests close to those of many feminists, especially those who regard logocentrism as inherently complicit with phallocentrism'.[24] If feminists wish to convert bodily experience from the reproductive 'weaker sex' to that of the reproductive and productive 'knowledge and power', I suggest we might need to work *from* our physical experience and find ways to articulate our embodied perceptions in the culture. We have to make sense of the world by thinking *from* our bodies in all their meanings. We need to write not only *on* our bodies – in the way that Hélène Cixous suggests is the desperate strategy of those who cannot put a verbal discourse into the culture (the hysteric) – but through, from and about them.[25] We might wrest the gaze from being *on* us to considering our own gaze on ourselves, not out of individualistic introspection but because it is a matter of necessity if we are to grapple with systems of control, not just in the media and politics but in science, technology and medicine. Women's bodies continue to provide spectacle for (professional) others. The gaze has now successfully penetrated our insides. The reproductive system can be converted into bio-medical imagery, whilst we lie supine and watch ourselves alongside our medical experts. We surely owe it to ourselves to see with our own eyes, understand with our own flesh, and know with our own bones? The sense of wonder digital technologies are bringing us is a new form of screen-based spectacle that merits vigilance. The artist Helen Chadwick, before her untimely death in 1996, became fascinated with the human embryo seen as it is, microscopically. In her last project before she died, *Stilled Lives*, impressed by how embryologists grade pre-embryos with the human eye, just as jewellers do, she accordingly arranged pre-embryos as brooches and necklaces, presenting these choreographed images hugely magnified. It was as if she was commenting on the absolute beauty of such early forms of human life, as well as how, through human manipulation, they become viable and, finally, how this is made a spectacle for the beholder. In this work the maternal body is absent, signalling how contemporary ART protocols disrupt the 'natural' embodied relationship between mother and embryo, and the extent to which such a separation leads to political and ethical controversy. In a commentary on these art works Louisa Buck considers that '[b]y bringing together these embryos and organizing them into "socialized" strings, clumps and clusters, she reinstates the notion of dependency whilst undermining the authoritative view of the foetus as a disconnected, solitary individual.'[26]

Stunning as this project is, it presents an unsettling and troubling paradox. By commenting on the objectification, commodification and fetishization of the embryo Chadwick actually reinforces the same.

The enthralling and spectacular beauty of the jewel-like pre-embryos she *arranges* for us, illuminated behind glass-like perspex, leaves us gazing at the orchestrated beauty of precious organic reproductive material. She makes us look at these images and their meaning, their provenance outside the mother, their status, and in so doing, framed as they are in the art gallery context, she re-renders them into publicly owned, cultural *objects*. With a final twist to the conundrum, these artworks have a price on them in the art marketplace.

Performance and ART

Given the statistic that one in six couples suffer involuntary childlessness and fertility problems in the United Kingdom it is striking how little such stories are told. But then, the sub-fertile tend to implode, be muted, suffering with gritted teeth out of a sense of powerlessness, failure, guilt and marginalization. Our voices rarely come through into the culture. It was such silence I chose to break. My more recent theatre research has reflected a subjective perspective in a triptych of very different projects on (non-) reproduction and correlative issues since 2000.

Yerma's Eggs (2003) is the second of these.[27] In this piece I aimed to confront the audience with the infertile experience and bio-ethics in an immediate, emotional and interactive way as only live theatre can do. It rejected conventional plot-driven narrative structures and we opted instead for a montage of spoken and unspoken scenes. If Lorca wanted theatre to be a passionate arena, a 'tribunal' for its audience, I wanted to explore how to get under the skin of the infertile subject, represent different cultural and sexual-choice perspectives and bring the bio-ethical debate on ART into theatrical space, emotionally, viscerally and inconclusively. Here there was no tragic ending, only miserable feelings wrestling with all the imperatives of medical intervention. *Yerma's Eggs* used the original play as a poetic framework, the plot filleted out so as to retain just the emotional flesh of barren grief. Yerma becomes every-infertile-woman (and man), all the actors playing her at different points and each exploring a position of infertility, desire and rejection by partner or community. In one scene a gay male couple discuss their ambivalent feelings towards parenting, in another (Lorca verbatim), Yerma and Juan fight in front of a massive projection of a five-week-old fetus, glowing like a crescent moon.

A multi-media approach and a multi-track montage narrative afforded a layering of meanings as well as permitting me to stretch the issues in historical time and geographical place. Video projection – *vox pop*, medical and thematic – could enter the action, becoming a protagonist in itself. We worked to find synergy between technology and action so that the projected material was, rather than illustrative or didactic, another layer in a stratified text. The technology of projecting images became a metaphor for collaboration with technology in ART treatments. Performers would nurse, move and embrace projectors so that film emerged from as well as on to their bodies. We projected on to flesh, the walls of the theatre, the floor, fabric, water and pouring sand. Images included 3D/4D ultrasound of the baby *in utero*, sperm entering egg, single egg cells, ICSI, sequences of children playing, spring flowers. Our finale included an empty swing going back and forth to the mournful sound of the creaking of its dry unoiled joints, a mechanism 'not put to proper use' as Yerma would say, and in need of attention.

The theme – our own concept of Nature, its cruelty, and where it might be abused by science – became a *leitmotif*. The intervention of medical technology in 'natural' conception is the *sine qua non* for the IVF user who is obliged to negotiate a range of chemical and technological paraphernalia.[28] Whilst doctors probe a woman's body with ultrasound, speculum and camera, society – provoked by the media – probes the very ethical rights we may have to interfere in Nature's choice. The public's disgust here – and confused ignorance – consistently returns to the same vexed question: is it *natural?* Legislation in the United Kingdom, upheld by the HFEA, has shifted this slippery notion of what is natural to focusing attention on two key issues: (a) the status of the embryo and (b) the welfare of the child. Our sentimental relationship with Nature and conversely suspicious one with science and technology reveal a profound and irrational idea of Creation itself. Despite living in a gizmo culture we are surprisingly circumspect when it comes to that which is manoeuvred (man-made) as it relates to making babies. ART consists of procuring cells by intervening in a natural process (egg and sperm collection) only to return the results of this (pre-embryos) to their natural environment (the womb). Attachment, the burrowing into the womb lining of such embryos so that they can be nourished by it, is never guaranteed. Indeed, problems of attachment, which so commonly lead to spontaneous or later miscarriage, remain one of the last challenges in IVF research. So babies born from ART might be the result of manoeuvred, technological pregnancies, but Nature, even now, still has the last word.

Discussions in the company regarding the role of Nature led me to write the performance's finale text. Against huge film projections of children at play, a performer holding a laptop bearing a film of a yellow flower in bloom asked the audience to contemplate what might be natural here:

Airtravel?
Fishfingers?
Ready to eat salad in a bag?
Sunscreen?
Coral?
Oxygen?
Mascara?
Condoms?
Aspirin?
Ice-cream?
Boats?
Olive oil?
IVF?
Lemons in winter?
The news?
Polystyrene cups?
Pubic hair?
Saffron?
Melodrama?
Families?
Talcum powder?
Sausages?
Sofas?
Nappies?
Sex?[29]

Yerma's Eggs was inevitably a highly charged, personal work bearing the stamp of each of my collaborators. The performer's body remained central, clothed and naked, screaming and whispering. The piece began with water, the source of life, and ended with the voice of my child. In between these lay the material of a performance that was inevitably uncomfortable for an audience. The subject is rife with contradictions, particularly for women. Splicing together a medical and technological world with the visceral elusiveness and tenderness of physical theatre posed many, but necessary challenges. To do justice to the subject matter, machine and flesh had to be married. Medical imperative on the one side and embodied emotion on the other are the dualities that the sub-fertile straddles, daily. We medicate and meditate. We inject and project. We imagine.

Spectacles

> Dark is dangerous. You can't see anything in the dark, you're afraid. Don't move, you might fall. Most of all, don't go into the forest. And so we have internalized our fear of the dark.
>
> (Cixous, 'The Laugh of the Medusa', p.336).

In the early 1970s Germaine Greer urged women to taste our menstrual blood, take possession of our pleasure, refuse what she named the 'Fear, Loathing and Disgust' that we internalized in self-hatred.[30] *Our Bodies,* as the American self-help book claimed, were *Our Selves* and we needed, urgently, to get *in* to them.[31] We enthusiastically squatted on mirrors and taught each other to use the speculum as a tool for empowering self-knowledge. Looking at our genitalia, this taboo-ridden mysterious unknown part of our anatomy, constituted an act of dis-covery, of revelation and reclamation. It was also, and importantly a flagrant protest against the way in which both society and the medical establishment had alienated us from ourselves.

Gray's Anatomy tells us that '[b]etween the clitoris and the entrance of the vagina is a triangular smooth surface, bounded on each side by the nymphae; this is the vestibule'.[32] Today's medical technology has taken us right into the house, deep inside our insides. Cordless ultrasound equipment (designed, for anatomical convenience, on a phallic principle) can be inserted into the vagina for the highest quality imaging, better than the normal scanner passed over the surface of the body part being surveyed (the womb). In the United Kingdom this remains a privilege for the fertility patient. The sight of your own ovaries, follicles or fetus represented by the fragile, ghostly black and white imagery of an ultrasonogram is astonishing. The shock is what in photography Barthes has called the 'punctum', the first impact of encountering an image, a 'sting, speck, cut, little hole...which pricks me (but also bruises me, is poignant to me)'.[33] It is a rare bonus of IVF to have frequent scans, beginning with the images of your embryos to late on in pregnancy.

Inevitably women develop a relationship with their embryos. The power of the image itself encourages bonding. Fervent attachment to the embryo is common in IVF, even in the atheist. It is a paradox that even among those who are politically pro-choice, when it comes to abortion they find a deep emotional connection to their pre-embryo(s). Perhaps such feelings can be better understood if, instead of seeking logical and ethical explanations, the embryo's meaning for the parent is considered. The embryo is the first success in IVF, a first achievement, and signifies a very real possibility of pregnancy and birth. Its image becomes a talisman, a pre-figurative icon of what might evolve to a baby. In ordinary reproductive lives the early embryo is unseen, in IVF it becomes a sight to behold. Imaging technology feeds the imagination, gives what would be invisible to the naked eye form and shape. It makes emotional connections at a cellular level, vividly and beguilingly. These 'embryos' are actually only a bundle of cells, perhaps eight on average, encoded with genes. In medical parlance they are 'pre-embryos' for they haven't reached the stage, 14 days after fertilization, known as 'primitive streak' when the first signs of the nervous system start to appear, at which point by law in the United Kingdom they have become embryos and cannot be used for research. The status of the embryo, the question of whether it is a person or a thing and at what point it achieves personhood remains the pivotal question in all bio-ethical considerations.

The theatre of technoscience

Sub-fertiles gain insight. We see infinitesimally small fragments of our genetic material grow. We are permitted this via available and ever-progressive imaging technologies. Becoming an IVF user is a journey into a collaborative relationship with science and technology. We mutate to cyborg, our chances of reproducing totally hooked up with and dependent on the application of technologies via which we see ourselves with fresh eyes, hope with fresh heart, submit our bodies to explicit interventions. Our reproductive strategies bond the organic to the inorganic, the programmed to the spontaneous, the predictable to the unpredictable. 'Cyborg replication is uncoupled from organic reproduction... By the late twentieth century, our time, a mythic time, we are all chimeras, theorized and fabricated hybrids of machine and organism: in short we are cyborgs.'[34]

ART poses questions about the very depths of our human identity. When the creation of life itself is dependent on all the stuff of medical and technological gear – chemicals, needles, surgery, incubators, cryopreservers, petrie dishes, laboratories, scanners and screens – we are negotiating a sometimes fraught and complex journey in which we permit ourselves to be manipulated, opened up, repaired, graded and exposed by machines – the non-human – so as to achieve the very human drama of parenting. Our bodies perform these dramas in ways which have become monitored visually, in and behind glass. And, as Haraway provocatively insists '[e]ach image is about the origin of life in a post-modern world'.[35] In what she calls the 'theatre of reproduction' and 'theatre of technoscience' we succumb to all the seductions, misconstructions and misnomers of the mediatized, visual culture. Furthermore, what is actually happening with reproductive imagery technologies is a return to Christian Creation narratives: 'It does not seem too much to claim that the biomedical, public fetus – given flesh by the high technology of visualization – is a sacred-secular incarnation, the material realization of the promise of life itself. Here is the fusion of art, science and creation. No wonder we look.' Haraway quotes the historian of the body Barbara Duden, who considers that the fetus has become a modern 'sacrum', an 'object in which the transcendent appears' (Haraway, 'The Virtual Speculum in the New World Order', pp.223–26). For Duden, the visual culture itself is what has conditioned our relationship to the naturally invisible, unborn, gestating life within the womb, so that our very idea of what the pre-embryo, embryo or fetus might be is inextricably linked to the picture we can have of it: 'The formation of the fetus is to a large extent the history of its vizualisation.'[36]

Hence the use (or abuse) of *Life* magazine photographer Lennart Nilsson's revolutionary images in his 1965 photo essay of the developing fetus *in utero* by anti-abortion lobbyists arguing for the rights of the unborn over that of the mother.[37] Paradoxically, these spectacular, aesthetically constructed, colour images – which Duden compares to nude photographs – were actually mainly corpses of unborn babies, removed from a dead woman and from a tubal surgery. Today, Nilsson's photographic techniques, published in glossy books since 1990,[38] have been surpassed. Mobile imagery of the unborn baby is made possible *in vivo* via 3D/4D ultrasound. It is currently possible on the internet to find clinics who, for a fee, will provide you with a video film of your unborn baby. The fetus performs.

Duden's work as a historian of the woman's body is inspirational. She writes, '[b]ody history, as I have come to recognise, is to a large extent a history of the unseen. Until very recently, the unborn, by definition, was one of these' (Duden, *Disembodying Women*, p. 8). Her thesis is that proprioception is culturally conditioned. How do we experience our bodies? How do we know ourselves as biological bodies and how does scientific knowledge about our biological bodies affect our proprioception? In turn how does this create and construct a language and self-descriptive narratives via which women 'present' themselves medically? If reproductive systems have become so spectacular, what are we looking *at* and how does this shift our idea of our body, our self and our role in reproductive processes? How can we repossess such imagery so as to promote our own, democratic interests in our complex and beguiling technoscientific reproductive future? Finally, biological imagery itself is mediated by the technologies used to produce it. So what are we actually looking at when we are presented, for example, with our pre-embryonic offspring on screen?[39] Duden calls such techniques 'misplaced concreteness' because they are images which could not be perceived by our senses. Images of a blastocyst are not only magnified but digitally devised, already processed for our eyes. I wish to draw attention to this, given the sense of wonder (Barthes's 'punctum') we all have when faced with brightly hued images of, for example, sperm working to crack into the surface of the egg.

We can see through flesh. We can view our bodies as inner spectacles *in vivo* where previously such visual opportunities, if they were to be offered the lay person at all, would be via paintings, sculptures, wax works.[40] The possibility of self-seeing has become (normally) painless and vivid. There are surely ever-increasing possibilities for empowerment in this, preceded as it has been by our earlier fumblings with the cold clamp of the speculum? The fetishization of women's bodies traditionally fixates on reproductive signs – the extruded breast and the intruded vagina. Fecundity doesn't feature in table dancing. Pornography hardly conjures ovaries, cervix, fallopian tube and womb. Haraway again asks us to consider that: '[t]he technologies of visualization recall the important cultural practice of hunting with the camera and the deeply predatory nature of a photographic consciousness. Sex, sexuality, and reproduction are central actors in the high-tech myth systems structuring our imaginations of personal and social possibility' (Haraway, *Simians, Cyborgs and Women*, p.169). Faced with the spectacle of our insides we surely cannot any longer remain a 'Dark Continent'? Our mystery is revealed. How will the depiction of our internal anatomies begin to shift the alluring promise that our external genitalia have hitherto offered the (male) for penetration? And how, conversely, will our sense of our own sexual signifiers shift in time as these visualization technologies journey us ever inwards to the phenomenal spectacle of our organic complexity? This explicitness, this overt and detailed imagery offered us via medical protocol, is doubtlessly going to affect our self-image, our sexual imaginations, but how we cannot yet chart.

As we celebrate the insights and knowledge that all such technologies might afford us, we need to remain cautious not to endow such images with false meanings, or meanings which disempower us by disembodying us. The danger is that as the fetus becomes more visible, so do we as women potentially become more invisible whilst, too, politicians and moralists might claim authority over our progeniture, as Pollack Petchesky[41] among others argues. We must remain vigilant in this new technological Eden, so that we might distinguish between the consumeristic notion of 'choice' provided by ART and the real meaning of reproductive freedom. Our reproductive future, whilst bearing the promise of freedom and choice that ART and reprogenetics might bring together with a potential revolution in sexual relations, will depend on our knowledge and our ability to challenge our legislators accordingly. For no amount of spectacle, insight or wonder provided by technology and no amount of successful IVF statistics can detract from the fact that where the real power lies is not with the tools of such technologies themselves, but with those who control them.

Notes

1 Jacqueline Brown, *Thinking Egg*, quoted in Deborah Law and Sandra Peaty, *Angels and Mechanics*, Catologue (London Arts Board/ Arts Council, 1996), p.47.

2 'At twenty week's gestation, the peak of the female's oogonial load, the fetus holds 6–7 million eggs. In the next twenty weeks of wombing, 4 million of those eggs will die, and by puberty all but 400,000 will have taken to the wing, without a squabble, without a peep' (Natalie Angier, *Woman: An Intimate Geography* (London: Virago Press, 1999), p.2).

3 In Vitro Fertilization. The *sine qua non* of IVF is that it constitutes fertilizing eggs outside of the body. Meaning, 'fertilization in glass', the IVF process actually takes place not in the ghoulish test-tubes of popular imagination but in a petrie dish. A great deal of detail goes into the preparation for this occurrence, with regular monitoring throughout. The procedure for the woman involves, first, a dose of hormones to shut down the menstrual cycle – in effect a chemically designed menopause – followed by intense doses of hormones taken subcutaneously to stimulate hyper ovulation. This causes the ovaries to swell (usually quite painfully). During this super ovulation phase you are ultrasound scanned, to check follicle and egg production. This is followed by egg-collection: surgery either via laparoscopy (keyhole surgery via an incision in the navel) or via the vaginal wall. Viable eggs can range from zero to the high twenties. Even more sometimes. If your eggs do fertilize, the pre-embryos are surgically transferred back into the womb or fallopian tubes.

4 Spit Lit Festival of Women's Writing presented by Alternative Arts, London, March 2005.

5 See Lee M. Silver, *Remaking Eden: Cloning, Genetic Engineering and the Future of Humankind*? (New York: Avon Books, 1997).

6 ICSI is the acronym for Intra-Cytoplasmic Sperm Injection, a relatively new technique used in cases of male sperm deficiencies, immotility or low counts for example.

7 See Dion Farquhar, '(M)Other Discourses', in G. Kirkup, L. Janes, K. Woodward, K. and F. Hovenden, eds, *The Gendered Cyborg* (London and New York: Routledge 2000), pp.209–220, pp.214–15; and Deborah Lynn Steinberg,

'Feminist Approaches to Science, Medicine and Technology', in *The Gendered Cyborg*, pp.193–208, pp.200–3.

8 Robin Baker, *Sex in the Future: Ancient Urges Meet Future Technology* (Basingstoke: Macmillan – now Palgrave Macmillan, 1999), pp.379–80.

9 See Phillippe Comar, *The Human Body: Image and Emotion* (London: Thames & Hudson, 1999), p.89.

10 Emily Martin, *The Woman in The Body: A Cultural Analysis of Reproduction* (Boston, MA: Beacon Press, 1992), p.20.

11 The Human Fertilisation and Embryology Authority, a government quango that licenses clinics and regulates legislation.

12 See Anna Furse, *Our Essential Infertility Companion* (London: Thorsons/ Harper Collins [1997], 2001).

13 For a brilliant analysis of Magdalene's role in Christ's life – and her possible pregnancy – see Elaine Pagels, *The Gnostic Gospels* (London: Penguin Books, 1990).

14 Rachel begs her husband Jacob for a surrogacy solution to her childlessness, asking him to go to her maid Bilbah and conceive their child with her.

15 Henrik Ibsen, *Hedda Gabler*, trans. Christopher Hampton (London: Samuel French, 1989), Act 3, p.50.

16 Federico García Lorca, *Yerma, Plays One*, trans. Peter Luke (London: Methuen, 2000), Act 1 Scene 2, pp.173–4.

17 Michelene Wandor, *AID Thy Neighbour* in *Five Plays* (London: Journeyman Press, 1984).

18 Michelene Wandor, *Wanted* (London: Playbooks, 1988).

19 Liz Lochhead, *Perfect Days* (London: Nick Hern Books, 1999).

20 Timberlake Wertenbaker, *The Break of Day* (London: Faber, 1995), Act 2, p.71.

21 Ben Elton, *Inconceivable* (New York: Bantam Doubleday Dell; new edition, 2000).

22 Adrienne Rich, *Of Woman Born: Motherhood as Experience and Institution* (London: Virago Books, 1977).

23 Maurice Merleau-Ponty, *The Phenomenology of Perception*, trans. Colin Smith, (London: Routledge, 2002), p.106.

24 Elizabeth Grosz, *Volatile Bodies: Towards a Corporeal Feminism* (Bloomington: Indiana University Press, 1994), p.94.

25 Hélène Cixous, 'The Laugh of the Medusa' [1975], in Elaine Marks and Isabelle de Courtivron, eds, *New French Feminisms* (Brighton: Harvester Press, 1981), pp.245–64.

26 Louisa Buck, 'Unnatural Selection', *Stilled Lives* (Edinburgh: Portfolio Gallery Catalogue, 1996).

27 I wrote and directed a play for 4–7 year olds, *The Peach Child* (Little Angel Theatre, 2000–1), based on an ancient Japanese folk-tale about a miracle baby boy born to a childless old peasant couple. *Yerma's Eggs* was a devised, physically driven, multi-media performance for a young person and adult audience (2003).

28 I wish to distinguish between science and technology. The word 'science' means 'knowledge' and 'technology' (from the Greek *techne*) means 'art' in the sense of skill, craft and how-to. To Heidegger technology is a 'bringing forth' that which is latent (Martin Heidegger, *The Question Concerning Technology and other Essays* (New York and London: Harper Row, 1977)).

Does ART (an apt acronym), then, merely bring forth the material possibility of babies from latent, organic materials (male and female gametes) which possess only potential life and only in combination with each other?

29 Abridged.

30 Germaine Greer, *The Female Eunuch* (London: Flamingo, 2003).

31 Boston Women's Collective, *Our Bodies, Ourselves: A Health Book for Women* (London: Allen Lane, 1978).

32 Henry Gray, *Anatomy* (London: Bounty Books, 1977), p.1027.

33 Roland Barthes, *Camera Lucida*, trans. Richard Howard (New York: Hill & Wang, 1981) p.27.

34 Donna Haraway, *Simians, Cyborgs and Women: The Reinvention of Nature* (London: Free Association, 1991) p.150.

35 Donna Haraway 'The Virtual Speculum in the New World Order', in Kirkup et al *The Gendered Cyborg*, pp.221–45.

36 Barbara Duden, *Disembodying Women, Perspectives on Pregnancy and the Unborn*, trans. Lee Hoinacki (Cambridge, MA: Harvard University Press, 1993), p.92.

37 See Duden, *Disembodying Women*, and Rosalind Pollack Petchesky 'Foetal Images:the Power of Visual Culture in the Politics of Reproduction', in Kirkup et al. *The Gendered Cyborg*, pp.171–92.

38 Lennart Nilsson and Lars Hamberger, *A Child is Born* (New York: Doubleday, 1990).

39 This idea lies at the centre of my installation *Photogenic Looks* at the Chelsea and Westminster Hospital, forthcoming 2006. With this work I am exploring the relationship between the body of the spectator and that of the disembodied reproductive material introduced to them. The purpose of the project is to narrow the distance between the image-object and its spectator. I want to convert the spectator, just as theatre can do, into participant. I want to engage the imagination of the spectator-participant, explore the emotions and physical responses the experience arouses and offer ways to locate, describe, analyse and express this verbally, visually, even physically.

40 Today, anatomy meets art. The magnificently curated *Spectacular Bodies* exhibition at London's Hayward Gallery in 2000, the hugely popular plastinated corpses of Professor Gunther von Hagen's *Bodyworlds* seen in London in 2003 and televised anatomy lessons on Channel 4 TV in 2005 are evidence that we retain a ravenous appetite for witnessing the explicit body, the sight of our bodies within, preserved, bottled, flayed or simply magnified.

41 See note 37.

11
'It is Good to Look at One's Own Shadow': A Women's International Theatre Festival and Questions for International Feminism

Elaine Aston, Gerry Harris and Lena Šimić

This essay is a three-way collaboration arising out of our participation in 'Roots in Transit', an international women's theatre festival held at Odin Theatret in Hostelbro, Denmark (15–25 January 2004). The three of us – Elaine, Gerry and Lena – work together in Lancaster University (United Kingdom): Elaine and Gerry as academic colleagues and as collaborators on a large-scale project concerned with women's contemporary performance (to which this collection of essays contributes); Lena as professional practitioner and research student (originally from Croatia, now living in the United Kingdom) working alongside Gerry and Elaine, participating in the large project and pursuing her own feminist performance research. Our interest in the 'Roots' Festival came out of these home-based projects, though we had different (feminist) ways into and different sets of questions and (theatre) concerns to bring to, the event.

In producing this essay we have attempted to preserve our different perspectives on this event. Functioning primarily as engaged observers and audience members at the performances, talks and discussion events of the second week of the Festival, Elaine and Gerry have melded their voices together to produce a more formal account: one that offers some theoretical reflections on issues of feminist theatre futures in an international context. As a participant in the first week of workshops as well as an audience member for the second week of events, Lena's interventions (presented in italics) are informal, more immediate and are deliberately presented so that they interrupt and lie alongside the more formal text, enabling the two to reflect on, comment on and even, at times, to contradict each other. Charged with the task of talking feminism, theatre and futures with other festival participants, Lena's interventions also bring in reflections from an international mix of women practitioners from different geographical, feminist and performance 'locations'. The punctuation and fragmentation of the formal text signals the 'imperfections' necessary to a 'futures' debate: figuring a desire to resist an homogenization of feminism, feminist theatre or feminist performance.

Roots in Transit

Roots in Transit was the fourth international women's theatre festival in a series inaugurated in 1992 by Julia Varley of Odin Theatret, Denmark. While Varley has lived and worked with Odin since 1976, she also has strong connections to the women's Magdalena Project dating back to the project's inception in Cardiff, Wales, 1986, and its First International Festival of Women in Contemporary Theatre. This first Magdalena festival seeded the idea of establishing an on-going women's network for an international exchange of performance cultures. Up until 1999, when it lost its Arts Council funding, the Magdalena project regularly presented workshops, performances and festivals. As the project successfully fostered links through associate or host projects, such as those with Odin and Varley, the idea of a 'space', an international meeting ground, for women practitioners to come together has succeeded in taking root elsewhere, with festivals in numerous places in the world.

What is the point of a network? Should I aspire to belong to one? Isn't it only natural that each network has its borders, issues and prejudices?

Like any other sort of 'meeting place', academic, artistic or political, theatre festivals are inevitably bound up in complex sets of relations, both material and discursive, that condition exactly 'who' is represented at them and in what terms. Yet this does not detract from the value and importance of the Magdalena project in bringing together work drawn from all kinds of theatre cultures and born out of different social experiences for women. Work that we saw at Transit included performances from, amongst others, artists based in Argentina, Peru, Spain, Italy, Bali, Morocco, Algeria, Egypt, France, Britain, Germany, Denmark and Taiwan.

A women's festival?

Many participants came with a previous interest and/or background in the Magdalena project while others were attracted not because it was a women's festival, but because it was held at Odin. It was noticeable, for example, that the festival attracted significant representation from countries where Odin's work has been influential through their festivals, work demonstrations, productions – especially parts of Northern and Eastern Europe and South America. Arguably, a proportion of the participants and the practitioners at Roots in Transit were already connected through shared *theatrical* 'languages' as much, if not more than, broadly *political* and/or specifically *feminist* interests.

Asked what it means to be a female working in theatre and a part of a women's international festival workshop, participants offered a wide array of views, ranging from disinterest in anything feminist (Cristina Galbiati, actress, director, and Ledwina Constantini, actress, from Switzerland; Aleksandra Ilić, visual artist, performer, Serbia and Montenegro) to political activism (Biljana Stanković, lesbian activist from Novi Sad, Serbia and Montenegro, within the organization Žene na delu – 'Act Women'.)

A number of people were affiliated to theatres or academic institutions that promote women's events, but do not declare themselves or their work feminist or gender specific. Quite a few participants felt deeply encouraged by being part of a female artistic community and a gathering that offered a nourishing environment, especially through contact with a large number of older women artists: Janica Draisma, filmmaker, director, performer, photographer, Netherlands; Karin Koob, performer, Germany; Laurelann Porter, playwright, singer, composer, United States; Margaret Cameron, writer, performer, pedagogue, Australia; Marisa Naspolini, actress, director, university lecturer, Brazil; Mirjana Vuković, actress, Serbia and Montenegro; Sharna Vrhowec, performer, Australia; Sandra Dempster, performer, Germany/New Zealand.

Many expressed a belief in a specifically female, nurturing, generous character when working in the arts, a somewhat maternal instinct that places emphasis on the process, transformation and making of the artistic material. Others such as Kasia Kazimierczuk believe feminism enhances a division between men and women that can be detrimental to art, but she acknowledges a need for a women's festival because a lot of female practitioners define themselves through gender and this festival is a way to reach them.

Janica, an artist from the Netherlands, observed that in her work she is interested in the struggle with male and female identities and their definitions rather than feminism.

Deirdre Roycroft, actress, Ireland (Loose Canon Theatre Company), came to the festival not believing it important that it was a women's festival, but left it feeling differently. In her opinion the work presented at the Festival was a kind of political activism, grounded politically not emotionally. She found herself thinking a lot about 'roots' – the festival's theme.

Feminism and theatre

We were drawn to this festival because our current research project is focused around women practitioners based in Britain and while unable, for pragmatic reasons, to extend its geographical scope, at the very least we wanted to extend our thinking around feminist futures by means of 'encountering' works from other countries as part of our process. As it turned out, from the very first night of the festival, we found ourselves being challenged to review our positioning on a number of levels. Not least we felt prompted to revisit the concept of 'feminist theatre' – something which, for various reasons, we have not specifically engaged with for some considerable time.

This is largely because recently, in the field we work in, the focus has been very much on particular modes of anti-essentialist theory. Within the terms of this theory, the question of what constitutes feminist theatre appears to have become redundant, inadmissible or answerable only in purely relativist terms. However, as the shows and discussions at Roots in Transit clearly demonstrated, the thinking which informs such conclusions is not necessarily shared by all women or all women theatre practitioners, on a 'local' let alone a global scale. In Europe, the United States and Australia the dominance of such paradigms has also sometimes led to the privileging of particular aesthetic strategies for political performance. As a result, certain types of women's theatre and performance-making have become problematic to discuss in 'feminist' terms. Within the academic field, this issue is sometimes avoided very simply through focusing on types of works that fit particular theoretical and aesthetic models, or by narrativizing shows through the dominant paradigms.

I fear to be known as a New Born Feminist among the participants of the festival. I make sure that the participants I talk to know I don't belong to the club.

It is not, we should stress, that we believe that there is or should be an answer to the question 'what is feminist theatre?', or even that it is the most relevant or important question to ask. Rather, our point is that attending this festival reactivated this question for us in a way that seems productive in challenging some of our *own* assumptions as feminist academics, concerned with theatre, theory and performance. Not least, this is because we were strongly reminded that these two words, 'feminist theatre' embrace (at least) four different, and complexly related, fields and different types of knowledges: feminist theories, women-centred social practices and experiences, theatre theories and theatre practices, all of which are also culturally inflected. How could any single 'theory' even begin to embrace the complexity of these fields and the way in which they touch on, complicate or cross each other's boundaries, *especially* in an international context?

The sorts of issues that arise from this complexity were demonstrated by audience responses to the very first show we saw. *Salt* by Odin theatre, was performed in Italian by Roberta Carreri, with musical accompaniment by Jan Feslev and direction by Eugenio Barba. *Salt* is described in the programme notes as 'A female Odyssey' in which 'A woman travels from one Greek island to another in search of a loved one who has disappeared. A phantom accompanies her in a dance, which brings her closer to an awareness of a definitive absence.' The programme notes anchor the show – but seeing the show without the notes (and without speaking Italian), as most of us did, all of us saw something different: it was about a woman's political life (she cradles a *red* book, while other images are reminiscent of torture/political protest); it was about the death of her children (lots of funereal images); it was about the pain of living (lamentational style of delivery), and so on.

At one level, we could argue that this is characteristic of theatre spectatorship in general: the old cliché of nobody seeing the same show. Aspects of *Salt* also suggested that it was a relatively 'open' poetic text, but this does not indicate a text that is essentially meaningless *except* in the process of individual interpretation. Moreover, in this instance, the variation between these interpretations and the programme summary specifically reflects the experience of watching a show in another language, even though for many of us the theatrical and cultural signposts were apparently familiar. How much greater, then, the problem with works in the festival where more marked theatrical and cultural differences were in operation, as well as linguistic ones. How as spectators and commentators could we begin to genuinely engage with these various 'differences', to do them justice without misrepresenting them or 'appropriating' them? In this context, on what basis could we speak about these shows in relation to feminism and theatre – in fact on what basis could we speak about them at *all*?

Encounters with theory

While these questions were prompted by the practice within the festival, paradoxically perhaps, we found ourselves turning to theory to help us work through these concerns. After all – we *are* academics. In the first instance, in our minds was Sara Ahmed's monograph *Strange Encounters*,[1] a work that is concerned with the problematics and possibilities of international feminism in the context of globalization. Writing in Britain and from the perspective of poststructuralist/ postmodern feminism, Ahmed is, nonetheless, as critical of the various ways 'difference' has been fetishized in anti-essentialist approaches, as she is of 'universalism'. Ahmed points to the way that such thinking can produce a return to 'universalism' by other means, analysing, for example, how anti-essentialist theories of cultural hybridity and diaspora can sometimes, as Ruth Frankenberg and Lata Mani argue, be used to imply that 'we can *all* [our emphasis] be defined as decentred, multiple, minor or metiza, in exactly comparable ways'.[2]

Ahmed argues that *both* essentialist and anti-essentialist modes of thought can assume to 'already know the difference', in ways that hold hierarchies in place and foreclose the possibility of reciprocity and negotiation (*Strange Encounters*, p.167). As such, her concern is to think through what might constitute 'ethical encounters' with different and distant 'others' in ways that avoid both the appropriations of cultural (and theoretical) imperialism and the 'indifference' of pure relativism. She stresses the importance of acknowledging the broader material and economic relations that inevitably inform such meetings, whether literal or figurative. However, by the same token she insists that they must be premised on a recognition of the *absence* of knowledge that would allow one to control the encounter or predict its outcomes, allowing for both surprise and conflict. In such encounters: 'one gets close enough to others to be touched by that [the different and specific histories and experiences] which cannot be simply got across.' In such encounters Ahmed argues, '"one" does not stay in place, nor does one stay safely at a distance (there is no space which is not implicated in the encounter)' (*Strange Encounters*, p.157).

In addition to Ahmed's thinking around 'strange encounters' we also turned to Eve Sedgwick's recent monograph *Touching Feeling*.[3] Through a discussion of affect, pedagogy and performativity, Sedgwick argues a need for thinking that helps to get beyond the versions of what we already know or what we have already learned to look for. Like Ahmed, Sedgwick is concerned with the way that anti-essentialist methodologies have hardened into an approach that has come to dominate, define and ultimately limit, the fields of enquiry in Euro-American feminism, queer theory and postcolonial discourse. She argues that these methodologies may 'have made it less rather than more possible to unpack the local, contingent relations between any given piece of knowledge and its narrative/ epistemological entailments for the seeker, knower, teller' (*Touching Feeling*, p.124). Yet Sedgwick's aim is not to critique or *reject* the theories that have generated this paradigm. Rather, it is to 'loosen them up', reopening their potential by considering them through other ideas and texts that might be excluded in advance by too rigorous 'theoretical hygiene'. One of the ways in which Sedgwick performs this is through drawing on Sylvan Tomkins's conceptualization of subjectivity, influenced by systems theory and based around 'theories of affect'. Sedgwick uses Tomkins's approach to argue that beside 'the structural dominance of monopolistic "strong theory", there may also be benefit in exploring the extremely varied, dynamic, and historically contingent ways that strong theoretical constructs interact with weak ones in the ecology of knowing' (*Touching Feeling*, p.145). As part of her argument she points out the 'interesting' nature of the proposition 'beside':

> a number of elements can lie alongside each other, although not an infinity of them .... Beside comprises a wide range of desiring, identifying, repelling, parcelling, differentiating, rivalling, leaning, twisting, mimicking, withdrawing, attracting, aggressing, warping and other relations.
>
> (*Touching Feeling*, p.8)

I am an artist and a student. I am female, and therefore I feel as though I am allowed to make feminism my calling card. Feminism gets me funded in my studies and in my art. In public I sell feminism.

On a more personal level I sometimes feel like an opportunist feminist, sometimes I believe in its causes, sometimes I find it silly, sometimes I am inspired by it, sometimes it fuels my anger, sometimes I forget about it...

I see feminism in academia as a safe friend to all the girls, in the art world I fear it is a quick categorization. There are prejudices about feminism; there are advantages as well.

Feminism is there to be used, abused, abducted, changed, born again, personalized – it is up to you and I.

To this we would add one further point of reference that informed our thinking around the festival experience: Janelle Reinelt's article on 'Staging the invisible'.[4] The link with Ahmed and Sedgwick is through the concept of 'affect', of embodied as well as abstract responses. Rooted in *theatre* as well as feminism (as we are ourselves), Reinelt, like Sedgwick, acknowledges the significance of anti-essentialist theory, whilst wanting to explore some of its limits. In particular, Reinelt takes issue with its potential erasure of local, contingent and material differences. Crucially, Reinelt inserts herself into her critical analysis as someone 'concerned with live performance', 'in a particular historical moment [that] seems infused with issues of identity that must be addressed' and that theory fails to touch ('Staging the invisible', p.99). Reinelt theorizes out of a theatre experience: a particular production of Caryl Churchill's *Cloud Nine* which, to her dismay, had recast the play's liberal take on sexual politics into a racist, sexist and homophobic discourse. It is an embodied, emotional response to this theatre production (she describes being 'deeply disturbed') that causes her to reflect back on theory and on theatre.

I would like to refer to Egyptian actress Maysa Zaki's words during one presentation/discussion at the Festival 'I am not Egypt.'

None of us is our country; nor the assumption of the portrayal of the country in the eyes of the others.

I am Croatian but I am representing England. I am the only workshop participant from England

We do not want to idealize Roots in Transit as an event. Nor in fact do we wish to repeat the, often formalist, romanticization of the potential for theatre and performance as a site of political transformation that has occurred in recent years in some feminist critical writing. Nevertheless, we would argue that at this festival, at least some of the theoretical concerns noted above were productively displaced. This was partly because, unlike some of the earlier Magdalena events that were premised on the idea of commonality (the idea that there might be a 'women's theatre language' for example),[5] the idea of 'roots' as a thematic for the festival did not suggest that the theatre works within it could, or should be, gathered together under a single banner. Rather, it encouraged an open dialogue between the women present around the concept of 'roots', theatrical, cultural, personal and political. Within this framework, for example, Hasna El Becharia, a Gnawa woman artist from Algeria, played songs from, or inspired by, the traditions of her people, including some performed on the gumbri, a sacred instrument usually strictly a male preserve. Cristina Wistari, originally from Italy and working with Ni Nyoman Candri, showed an example of the Balinese form Topeng Shakti, historically another male genre of performance. Teresa Ralli, a founder member of Grupo Cultural Yuachkiani from Peru, performed a one-woman adaptation of Sophocles's *Antigone*. The Uhan Shii Theatre Group from Taiwan performed *My Journey*, based on the life of Yue-Sha Shei, a Tawian Opera actress. Claudia Contin offered an interpretation of Arlechinno very much 'rooted' in the history of this figure within Western popular theatre but also filtered through her training in other performance techniques such as Kathkali. On some level, and in different ways, *all* of these performances were interventions within and rearticulations of various cultural and theatrical traditions, and this was often the basis of their identification with some sort of feminism. Yet many of these artists also spoke passionately about their 'rootedness' within some of these traditions, and of the desire to preserve and disseminate them, whether they were 'adopted', borrowed or indigenous. This desire was sometimes separate from and took priority over a concern for gender. What emerged from this dialogue, then, was a wide variety of relationships to, and understandings of, the idea of 'roots' in relation to both feminism and theatre that embraced 'desiring, identifying, repelling, parcelling, differentiating, rivalling, leaning, twisting, mimicking, withdrawing, attracting, aggressing, warping' (*Touching Feeling*, p.8).

Listen, learn, labour

In this scheme of things, our 'ecology of knowing', our dominant view or expectations of feminism and theatre clearly need to be transformed through a way of 'seeing' or knowing that is arguably less certain, more diffuse, much looser, the acknowledgement of something much 'messier' than allowed for by strict 'anti-essentialist theoretical hygiene'. Yet equally, it is important that this diffusion or messiness does not become simple relativism, a proliferation of differences that, to borrow from Judith Butler, 'leaves no means of negotiating between them'.[6] We would argue that this was avoided at the festival by the common and deeply pragmatic focus on the *work* of theatre, creating a space where a wide, but *not* infinite, variety of relations towards 'roots' could lie alongside one another.

Being a Croatian I find it the easiest to hang out with the Serbians, ironically...my war enemies. We share the language, and I do not have to think about what is polite/ok to say. I guess we allow each other in language much more freedom of expression, we are not guarded.

We women stayed in the Holstebro Sports Centre: four women in a room, eight women sharing a bathroom and a toilet. We were brushing our teeth, having a wee and a shower in front of one another. The only thing we didn't do collectively was poo. We were all fed up with one another after a while. It was just too intimate and claustrophobic.

Sharing the room with an Australian, a Brazilian and a German I realize that how I talk to them, how I dare talk to them, depends on my perception of their respective cultures. The more I know (or think I know) about the culture the freer I am. I end up having a laugh with Sharna from Australia, maybe on the assumption that she must be similar to my Australian friend Sonya. With Marisa (Brazilian) I can't get past rather formal sentence structures. We actually talk about this problem and she assures me she has such a laugh with Laurelann from the United States because they speak in Portuguese. 'Language is all', she concludes.

I wonder whether it is all about respecting each others' cultures. We are so removed from one another that we dare not have a conversation about it. We all hang on to our PC positions. We believe we can only learn from one another. We can listen. We cannot comment. We have no knowledge, we are not experts. The cultural difference is greater than theatrical likeness or rather, we fear the cultural difference more than we trust the theatrical likeness.

In her opening remarks Julia Varley encouraged us all to be active participants in all aspects of a programme in which the day usually ran from 9a.m. to at least 11p.m. and included up to five shows, workshop demonstrations, and talks and discussions at which practitioners offered autobiographical stories about both their personal lives and professional practices. These discussion spaces were constantly filled with the murmurs of many voices as clusters of participants translated for each other, often by means of a 'chain' that could pass through three different languages. All participants also had a commitment to helping the production teams and to the cleaning or housekeeping of the festival community. In part this reflects Odin's own style of 'third theatre', but in the context of a women's theatre event the democratization of domestic labour takes on a particular meaning (given the various histories of women's domestic oppression and exploitation). In her conclusion, when Ahmed models a 'collective politics' that avoids the violence of 'we' (the hierarchical 'we' of a Western definition of first-third-world feminism, whether essentialist *or* anti-essentialist), it is one which places a particular emphasis on labour: the 'painstaking labour' of getting closer to each other, working for each other and speaking to (not for) others in order to find out what as women 'we might yet have in common' (*Strange Encounters*, p.180). Getting closer is not a given, nor easily achieved. It is something that 'we' all have to work for and something we all have to be 'in'. We would argue that this context, the emphasis on the 'sharing of work', on working *for each other*, also foregrounded the 'ethical' responsibilities and the 'labour' involved in listening and spectating, as well as speaking and performing. In these terms, for us as feminist theatre academics, then, this space activated us to question how one can bear witness 'without presuming that such witnessing is the presenting or ownership of "the truth"' (*Strange Encounters*, p.158).

I appreciated the performances while observing them with the eyes of a theatre practitioner; the effect they had on me was both cultural and artistic. The ones from Argentina, Bali, India, Peru, Egypt, Taiwan (including the concerts from Algeria and Morocco) I watched with cultural distance and respect, while playing with the question of could I (ab)use some of their elements in my work...I guess I couldn't help activating my neo-imperialistic gaze. I have lived in Britain too long.

The other ones, those that come from cultures that are more familiar to me, like Denmark, Spain, Wales, Australia, Norway, France, Germany, Italy, I scanned for the artistry and the use of theatre language while at the same time I let myself indulge in the emotional and intellectual realm of the effect/issue the performances gave rise to.

It is Good to Look at One's Own Shadow

While, as indicated above, these issues were immediately 'opened' at the festival by *Salt* they were brought into sharp focus by other shows, such as *It is Good to Look at One's Own Shadow* written directed and performed by Luisa Calcumil from Argentina, also appearing on the first night of the festival. By the very nature of its subject matter, *Shadow* raised complex issues of appropriation and 'colonization' of works from other cultures, whether in the process of academic interpretation or by other means. Created by Calcumil in 1987, and remaining in her repertoire by public demand, this was the first time that *Shadow* had been performed outside of Argentina. According to the programme notes, core to Calcumil's performance is the idea that looking back 'at one's own shadow' is a way to 'move forward and to allow us all to rethink our principles of justice love and freedom'. Briefly, the piece is concerned with white cultural and economic imperialism, the oppression of indigenous Mapuche Indians, the destruction and loss of their traditional lifestyle and culture, and their continuing positioning as marginalized 'others'. It is composed in three parts, around three characters: a Mapuche woman who encounters the violence of colonization; an elderly woman trying to keep alive the memory of Mapuche culture; and a young woman (her daughter) living in a city where her experiences of exploitation, degradation and loss of 'rooted' identity are driving her to the brink of madness.

As Luisa later confirmed, *Shadow* was designed as a touring piece, to be performed in a wide variety of sites and without the technical resources of a theatre building. As a result, its staging was simple, with sticks, ribbons, bits of material, some fencing and plastic bags being used to define the performance area and to suggest different places and periods, and at one point a papier mâché mask was used to represent the forces of white colonialism. Alongside this broad stylization, however, Calcumil offered a series of stunning, technically accomplished 'naturalistic' transformations from the mother to the old woman to the young girl.

The programme notes suggest that Calcumil operates from an 'essentialist' understanding of cultural and gendered identity, at odds with the 'theories' that have recently dominated white Anglophone academia. Even more importantly, in the first section vocalization (mainly song) was in the Mapuche language, Mapudungun, and the rest of *Shadow* was in Spanish. Our knowledge of both Mapuche culture and of Mapudungun is non-existent, and our Spanish extremely limited, so there were clearly many layers of 'meaning' simply not accessible to us. Could we speak about this show without 'misrepresenting it' – ethically, if we speak at all, surely we should simply emphasize our 'distance' from it?

Yet, as Sara Ahmed argues, there is every need to talk *because* we don't speak the same language (*Strange Encounters*, p.180 our emphasis). Indeed, whatever was 'lost' in our viewing, the urgency of Luisa's desire to talk, to engage in an exchange with as many 'others' as possible, was clear both in this show and in (translated) conversations with her. Moreover, in these cross-cultural women's theatre contexts, our semiotic activity as spectators is problematic if it serves to create what Ahmed describes as a 'new "community of strangers"' (p.6). In brief, if we keep ('read') some bodies as *already* distant to us, then we are in danger of asserting our own identity and cultural authority whilst positioning 'others' as alien. This, Ahmed suggests, has, for example, been the problem in Western feminism's positioning of 'third-world' women. Drawing on Chandra Talpade Mohanty, Ahmed argues: 'third world women come to define not simply what Western women are not (and hence what they are), but also what *they once were*, before feminism allowed Western women to be emancipated' (*Strange Encounters*, p.165).

This allows Western feminism to adopt its 'superior' position: the 'ideal' for other pre-feminist, 'primitive' societies to aspire to. It is the means by which a global 'us and them' binary is kept in place.

In fact, we could (apparently) bring *Shadow* 'closer' to 'our' position by narrativizing it in relation to (anti-essentialist) notions of the cultural hybridity of colonized subjects. Indeed, the performance *does* show evidence of such hybridity, but this actually says nothing about the show and could be assumed in advance, without actually seeing it. It would also constitute an assertion of our own authority at the expense of Calcumil's. In short, to privilege such a reading is potentially to enact another form of 'colonialization' that detracts not only from our own lack of cultural and linguistic knowledge but also from the 'local' specificity of the experience Calcumil works so hard to communicate. Above all, it completely belies what moved us in this *theatrical* encounter, which works not only through the 'abstract and cognitive' but also through the personal, the embodied and emotional, so that distance works 'beside' identification. Indeed, within the show we felt 'addressed' so as to make an immediate identification with Calcumil both as a performer and as a marginalized subject. Yet at the same time 'we' also felt addressed as subjects in and of history, encouraged to look at our 'shadow' in the guise of the ugly, comic mask of the arrogant and violent imperialist. In short, this passionate, political work of theatre demanded that we get closer, to 'be in it', so as to recognize 'the responsibility the self has for the other to whom one is listening' (Ahmed, *Strange Encounters*, p.157). We cannot presume to present or 'own' the 'truth' of this show, but we can attest to its affect, the way it 'touched' us, in both Ahmed's and Sedgwick's sense, making us aware of 'histories, traumas, scars, wounds', which 'cannot be simply got across' (*Strange Encounters*, pp.157–8).

My Journey

As a theatrical encounter, what we are trying to articulate about *Shadow* could possibly be marked down as a response to its 'form', which bore certain similarities to structures employed within political community works that we saw in Britain at the time of its original creation. However, we would argue that the politics of a performance can never be reduced to its form, unless certain forms are assumed to be 'essentially' (even universally) reflective of particular meanings.

This has sometimes been a danger in the way that the concept of gender as performative has been taken up and applied to theatre practice.

At this festival *My Journey*, performed by Uhan Shii Theatre Group from Taiwan under the direction of Ya-Ling Peng, cried out to be interpreted through this theory. This show was based on the life of Yue-Sha Shei, one of Taiwan's most celebrated Taiwan Opera performers, now in her sixties. In Taiwan Opera musicians are always male, but performers are exclusively female and Shei specialized in male roles. *My Journey* is actually performed by Shei with two other female performers, who enact the conflict she experienced between the 'two voices' inside her body, as indicated in the programme: 'One said I am a very handsome man/ One said feminine charm is my nature', and finally her realization that 'I have many facets I can be handsome like a man and charming like a woman.' This story is played out using the form and techniques of Taiwan Opera, with its elaborate and colourful costumes and highly codified gesture, dance, song and music.

There is no question that within this show, to 'us' at least, gender appears to be marked as a series of Butlerian 'stylized bodily acts'. Yet to interpret this show in this manner feels like a failure to 'listen' and a reduction and oversimplification. This is not just that it potentially ignores the immensely complex and violent history of the Taiwanese context from which this show, and the form of Taiwanese Opera in general, emerges. Indicated by Peng in her talk earlier on the day of this performance, this includes the legacies of various occupations of this island by the Dutch, Japanese and Chinese Nationalists, as well as post-war interventions into its political status by the United Nations and North America. As a result of this history, Peng indicated the existence of deep-rooted ethnic and class divisions and a contemporary drive to 'modernization', on a distinctly 'Western' model, all of which colours indigenous attitudes to Taiwanese popular forms.

Nor is our concern entirely a matter that, in this instance, we lack the information, or the tools, to identify the model of subjectivity and identity ascribed to by the show's creators. Both these points, once again, mark our inability to fully 'read' a show, which is also in a language and, as importantly, a performance tradition to which we have no access, except through the sheer theatrical pleasure offered by the dazzling beauty and charm of its staging, a pleasure rendered suspect by the spectre of 'orientalism', of fetishization of the exotic 'other'.

Sedgwick argues that the sort of approaches that inform Butler's theory of gender performativity have come to function as a 'strong theory of affect' characterized by paranoia. As a strategy it seeks to forestall pain and prevent being surprised by the negative affects of violence and oppression, through means of foregrounding, revealing, exposure and demystification. In theatrical terms, both gender performativity and some conceptually related modes of postcolonial theory have come to be associated with an aesthetic that depends on distancing effects, usually through the use of mimicry based on irony, parody and pastiche. As already noted, Sedgwick points out that this type of 'strong' theory may be wide reaching but can also ultimately be self-defeating, finding only what it was 'looking for in the first place'. She also suggests that an overemphasis on this paranoid position overlooks the potential of 'reparative' strategies, which Melanie Klein relates to the depressive position. In such strategies one uses one's own resources to 'reassemble' or 'repair' the part-objects that induce anxiety into something like a whole, though she emphasizes '*not necessarily like any pre-existing whole*' (Sedgwick, *Touching Feeling*, p.128). As Sedgwick points out, the reparative is deeply engaged in the aesthetic and one of the words Klein uses for the reparative process is 'love'.

It seemed to us that the aesthetic in play in *My Journey* was far closer to something like a 'reparative' strategy than a 'paranoid' one, in its relationship not only to its 'roots' in Taiwanese culture but also towards gender and to its audience. Peng's work in recent years has been with a form of 'community' theatre working with the elderly and *My Journey* appeared committed to being 'accessible' rather than 'knowing', drawing the audience close rather than creating distance. In this festival, this included providing enough information in the programme to follow the story and, in the course of enacting the conflict between Shei's 'two facets', the show itself also demonstrated and 'explained' some of the performance techniques and vocabularies relating to different roles in the Taiwan Opera. Within these scenes, there is evidence of humour and even perhaps the most gentle play of irony around arguments as to whether performing male or female roles was more 'skilful' but there was no sense of distance, either from the traditions of Taiwan Opera or these roles. This is because the 'troubling of identity' performed in this piece is entailed with Shei's *labour* as a performer in this tradition, which started with her training at the age of five. It is this 'life' lived out in this theatre, in which performers are allocated male or female roles at a young age, that produced the 'conflict' staged in this piece. Yet *My Journey* appears as a celebration of this life's work, an act of love for this theatre form, using its resources to repair the anxieties it has induced into a 'whole', by means of telling a story that is wholly theatrical and yet also deeply felt and profoundly personal. This 'whole' is not one that can be simply used to 'exemplify' abstract (Western) social constructionist theories of gender performativity not least because this is marked as *Shei's journey*, rooted in her very particular embodied and located experience. And this experience both is and is not represented by this theatre piece. The audience is invited to get up close and personal, to encounter her 'voice(s)' in a way that works against the distance required by 'exoticization', but her 'voice(s)' cannot be detached from the highly stylized, culturally specific nature of this form. Within this form Shei's 'offstage' life and experiences remain unperformed, not available to appropriation, yet nonetheless not elsewhere, haunting and informing the piece. It is what makes it 'moving', it is what moved us, preventing us from identifying her or this theatre form as a distant 'other', yet equally from asserting our own cultural authority.

The return of the maternal as feminist subject?

I was especially interested in Gilla Cremer's performance of m.e.d.e.a., *because at the time I myself was working on* Medea/Mothers' Clothes. *This was a live art event in which I was refiguring* Medea *through the local and the lived experience of being a foreigner and a new mother in Liverpool, United Kingdom. I came to Cremer's performance, then, through my own reading of the* Medea *'text'.*

As the punctuated, broken up title suggests, m.e.d.e.a. *offers no unity. The narrative and the style of acting are fragmented and Cremer uses elements from different* Medea *texts (Euripides, Hans Henny Jahnn, Heiner Müller, Pier Paolo Pasolini) and her own text about a contemporary woman, Renate. Renate has been left by her partner (for a younger woman) after 20 years of cohabiting. Cremer also draws on autobiographical material for this piece and* Medea *is 'recycled' into a text focusing on a generation of German women who were 'young and crazy' in the 1970s. Renate is shown as having been betrayed not only by her partner but also by the promise of liberation in this past. Cremer and her director are interested in looking at how contemporary divorced or separated women deal with feelings of rage, pain and desire for revenge.*

m.e.d.e.a. *'reads' Medea specifically for its theme of betrayal while my own* Medea/Mothers' Clothes *is preoccupied with the themes of foreignness and motherhood. There is no single or correct reading of* Medea, *it is a text that is very open to interpretation. One can play with* Medea, *and the archetypal images she has come to represent.* Medea, *primarily a patriarchal product, is often read as the ultimate rebellious figure. In feminist hands, she is used as a tool and transformed into an anomaly, capable of invoking fear and horror in a way that destabilizes contemporary social structures.*

However, each feminist intervention into the Medea *text is read within the borders/roots of its context during its inception and reception. Gilla Cremer invented Renate, socially rooted in what I assume to be a German middle-class context for a specific audience in Germany. However, for the Festival's international audience Renate speaks English. The language element of the performance is modified by being 'universalized' – English is used in its 'International Headway mode'. Renate loses her regional and social grounding, while keeping the characteristics of her specific situation. The cultural transference within the frame of an international women's theatre festival produces a (non-)specificity of character – German, and yet not German, socially belonging somewhere and yet aloof – and for me this makes the performance 'rootless'. Yet between my own 'rooted' perceptions about* Medea *as a spectator and Cremer's as the maker of the piece, for me,* m.e.d.e.a. *makes sense of the festival theme, demonstrating how 'roots' can be 'in transit'.*

Obviously we do not have space to discuss *all* the shows in this festival either individually or in terms of identifying the interplay of connections and difference between them. However, in concluding we want to touch on one final issue the festival raised for us, specifically the way that the recent wave of anti-essentialist 'theoretical hygiene' has made the maternal a particularly difficult feminist subject. Many of the festival shows focused on the personal, autobiographical stories of mothers. For example, in *Looking for the Meaning* Gilly Adams (Wales) chose to focus on the experience of her mother's recent death as a way of also narrating her own experiences of ageing. Described as 'a way of keeping memories alive', Iben Nagel Rasmussen (from Odin) performed *Ester's Book*, a powerful work in progress that looks back over and stages the life of Rasmussen's elderly mother now suffering from dementia. As spectators coming from our particular generation of Western feminism, we are mindful of the essentialist difficulties an abundance of mother figures suggests, and cautious about the dangers of reclaiming the mother as some kind of celebratory, universal figure for feminism. However, being *moved* to think about the maternal because of these shows, and 'ungrasp[ing] [our] hold on some [maternal] truths that used to be self-evident' (Sedgwick, *Touching Feeling*, p.3), allowed for the possibility of encountering the maternal as potentially crucial to feminism in a global (futures) context. We would like to end therefore by considering this issue in relation to Anna Yen's *Chinese Take Away*, performed in extract at the festival by Yen and screened in its full cinematic version.[7]

Chinese Take Away tells the story of Anna Yen's Chinese mother who, in the 1950s, was coerced, tricked by her family into leaving Hong Kong for an arranged marriage in Australia. Torn from her Chinese roots, a victim of rape, domestic and racial abuse, her mother committed suicide when Yen was in her teens. In stage and film versions of *Chinese Take Away* Yen's style is presentational: she plays and presents all the family characters in her story (including her father). *Chinese Take Away* arguably belongs to an autobiographical tradition of solo work that draws on a feminist legacy of the personal as political: stages one woman's personal story in an epic, political framework. In piecing together the story of her mother's life, her enforced marriage, her rape and her suicide, Yen also connects to her grandmother's stories of being sold as a child into servitude. The show is a constant weave between the autobiographical (as Yen comes to an understanding about her mother's suicide and learns more about her family) and the political (as she exposes the mother's suicide as consequent upon and resulting from a damaging set of patriarchal, colonialist and material conditions).

As spectators of *Chinese Take Away* the mother's story is one that affected us emotionally: prompted connections to our own mothers (one living and one dead) and to ourselves as daughters. It was clear at the end of the film screening that many, if not a majority of the mostly women spectators were affected, moved in similar ways. To make an observation of this kind, however, runs contrary to the 'anti-essentialist project' informing recent feminist thinking. To state the powerful connections women have to the mother figure seems like a reductive return to the limitations set by an earlier mode of cultural-feminist, essentialist thinking: one that risked fetishizing the mother figure and prioritizing the mother–daughter 'bonding'. If we 'loosen up' on the 'anti-essentialist' feminist project, and if we acknowledge the maternal feelings stirred up in this viewing, can these be helpful to us in some other (less reductive) ways? And is it possible to 'stir up' our theoretical response? What comes into view if we risk this 'step to the side of antiessentialism' (Sedgwick, *Touching Feeling*, p.6)? Can it 'move' us towards another way of *knowing* the maternal? Harder, 'messier', but arguably more progressive, we would argue, are the (future) ways of 'surprising' or of 'moving' our own Western feminist thinking through the maternal to a different 'place' of knowing.

Crucial to this shift in knowing, is the way in which Yen's *personal* story about her mother is presented through and as a part of an epic history of patriarchal and colonialist violence. As spectators, when we are touched by Yen's personal portrait of the maternal and feel our own maternal connections to this, *our connections are also being disturbed, moved by, this colonialist history* – one that is not our own, is distant to our own, but that we are 'touched by'. It would be easy, too easy, dangerous even, for us to 'take away', to assimilate, Yen's imaging of Asian-Australian maternal ancestry back into the 'mother' as she is already known to us, both personally and theoretically. Yen signals this danger in *Chinese Take Away*: it is the mother's assimilation into white Australia that covers up the violent domestic, cultural and social history that she is subjected to. If we want our feminist thinking to move to a different place, we must avoid the (colonialist) temptation to think we already know this 'distant' mother by taking her for our 'own', and, at the same time, allow her to 'mess up' our own personal and feminist histories of feeling and knowing the maternal. If we allow for this to happen, for example, if we move closer to the colonialist history Yen portrays, then 'seeing' this, knowing this, we cannot simply, easily, go back to the maternal as figured in European feminisms (from Kristeva, Cixous and others): as a site of marginal-to-be-celebrated-potentially-transgressive-otherness (not least because in psychoanalytic terms the colonialist history forecloses on the polymorphous *jouissance* of mother–child relations). Nor, given Yen's representation of this damaged maternal in both the psychic and the social, can we resign, abject the maternal to the margins of a feminism that has bordered its thinking through its insistence on an anti-essentialist position. Rather, the maternal emerges as something women have in common (we all have mothers at some point, we all have feelings, different feelings about mothers), embodied responses, but not to be read back into a reductive site of essentialist, universalizing celebration (of what we have in common), but in view of the local specificities of cultural, colonialist, psychic and social damage(s) emerges as an issue that demands an urgent (future) critical attention.

The final image of *Chinese Take Away* is of Yen swinging backwards and forward on a slackrope. In the screen version she is suspended above the land: swinging backwards and forwards, facing out to sea. The image suggests the idea that histories, ancestries, are balanced by, are in tension with, the promise of future directions. It is important, however, that the figure of the daughter, Anna Yen, now performing herself, is in movement. Representationally, this resists the idea that Anna is somehow 'complete': has found out what she needs to know and can move forward leaving her past behind. Instead, it suggests a constant, perpetual movement: forwards and backwards, forwards and backwards, in tensions of knowing and not knowing, constantly transforming.

Similarly, the combined affect on us of these various theatrical 'encounters' in the space of the festival has been to move us to look back at our own feminist 'shadows' and forwards to future ways of knowing. Equally, as an experiment the creation of this dialogue between three voices has itself been one of movement, forwards and backwards between us, a process that has not been without its tensions, surprises and conflicts. Despite its moments of apparently authoritative (and theoretical) statement, this process is not at an end. As three different individuals interested in feminism, theatre and performance, we are still at the start of using the space opened up and inhabited by this festival, to find out what as women we might *as yet* have in common.

I think of Mirjana Vuković, a Women's Studies graduate and a performer from Serbia and Montenegro. Because Mirjana graduated in Women's Studies she doesn't have a formal theatre education (which is a must if one is to do professional theatre in the Central/Eastern European countries, because the whole of the professional and academic theatre culture tends to be centralized around one Academy of Dramatic Arts). But Mirjana came to theatre during the bombardment of Belgrade by participating in some drama workshops. She found that theatrical space offered the possibility of being inside and outside at the same time. It is her longing to be outside in the street during the danger that brought her to the safe inside of the theatre. With two other women she founded a collective 'Feministička pozorište' ('Feminist Theatre'). They decided to call it 'feminist' in order to create a polemic, to provoke discussion. The group abolished hierarchy in favour of more tolerant structures, but found this difficult, encountered conflicts through a lack of clear aims, and fell apart after a couple of years.

Yet Mirjana is moved and surprised by women's creative and political efforts and energies. She feels that in Serbia and Montenegro it was the women who took charge of the peace process and saved the disintegrating academia during the years Serbia and Montenegro were facing sanctions and cultural and political isolation from the rest of the world. Women, she contends, are also a moving force in the theatre scene in her country with Dah Theatre, Omen Theatre, Ivana Vujić and Biljana Srbljanović. Now Mirjana is thinking about theatre through female motifs, about exploring a new aesthetic and what that might open up in and for the future.

Notes

1 Sarah Ahmed, *Strange Encounters* (London: Routledge, 2000).
2 Ruth Frankenberg and Lata Mani, 'Crosscurrents, Cross Talk: Race "Post-Coloniality" and the Politics of Location', in Kum-Kum Bhavnani, ed., *Feminism and Race* (Oxford: Oxford University Press, 2001), pp.479–516, p.487.

3 Eve Kosofsky Sedgwick, *Touching Feeling: Affect, Pedagogy, Performativity* (Durham, NC: Duke University Press, 2003).
4 Janelle Reinelt, 'Staging the invisible: The crisis of visibility in theatrical representation', *Text and Performance Quarterly*, 14: 2 (1994): 97–107.
5 Susan Bassnett, *Magdalena: International Women's Experimental Theatre* (Oxford: Berg, 1989) writes: 'Magdalena had started with an idealistic, some would say utopian, belief in the capacity of women to come together and share their lives and work. Differences of culture, language, race, class, age and sexual propensity would, in this idealistic vision, be subordinate to the commonality of women's experience . . . (p.94).
6 Judith Butler, *Bodies that Matter: On the Discursive Limits of 'Sex'* (London: Routledge, 1993), p.114.
7 Anna Yen performed the opening to *Chinese Take Away*. This was immediately followed by a screening of the film version.

12
Gendering Space: The Desert and the Psyche in Contemporary Australian Theatre

Joanne Tompkins

> Geography and space are always gendered, always raced, always economical and always sexual.[1]

A discussion of spatiality in a collection on feminist theatre and its future(s) is as essential as a discussion of bodies or sexuality or any other major concern of/for women, since the function of space/place directly influences individual identity construction:

> Identity, as well as being about identification and organisation is also about spatiality. In part, this means that identity involves an identification with particular places, whether local or national. It also means that certain spaces act as sites for the performance of identity.[2]

Further, space and geography also affect identity construction at the level of cultural organization:

> Just as none of us is outside or beyond geography, none of us is completely free from the struggle over geography. That struggle is complex and interesting because it is not only about soldiers and cannons but also about ideas, about forms, about images and imaginings.[3]

In these terms – and most especially in the context of the urgent need to address the material and psychic effects of the geopolitical reconfigurations produced by the processes of globalization, past and present – a concerted focus on the relationship between subjectivity, space and spatiality must necessarily be part of feminist thinking about the future, both political and theatrical.

In this chapter I explore these issues through an examination of the ways that four contemporary women playwrights have found the Australian desert a fertile setting for explorations of landscape, corporeality and subjectivity. While theatre has not completely overlooked the desert in the past, its use as a location has been more common in other genres, notably film and fiction.[4] The renewed interest in desert locations that characterizes the four plays addressed here would not surprise Roslynn Haynes, whose analysis of desert representations in Australia illustrates that:

> few other landscapes have been so variously perceived or have elicited such diverse responses as the Australian desert. In the two centuries since European settlement of the continent it has been promoted from 'best forgotten' oblivion to centre stage prominence. Once execrated for its failure to provide an inland sea, the Centre is now the most exported image of Australia: in tourist offices around the world Uluru vies with the Sydney Opera House as *the* icon of the continent.
>
> (Haynes, *Seeking the Centre*, p.3)

Often called the 'dead heart,' the extensive landscape that could be termed desert in Australia is remote from most of the nation's inhabitants who live on the more verdant perimeter of the continent. Plays set in the desert naturally explore this remoteness, but there is, of course, more to the use of this location. The relationship between (white) women and the desert is, according to Rachel Fensham, 'an encounter with the unknown, here not only the feminine unconscious, but also the "political unconscious" of the nation. And it is a place in which ghosts appear, shadowy figures, unfamiliar to the urban fringe dweller.'[5]

Therefore, the desert setting in Australian plays by women operates on at least two levels: it fixes a literal, geographical site on the map, and it also marks the significance of politically important psychic 'locations' that may be less identifiable on a conventional map.

These tropes are very much in evidence in Jenny Kemp's *Still Angela* (2002), Noëlle Janaczewska's *Historia* (1996), Josie Ningali Lawford, Angela Chaplin and Robyn Archer's *Ningali* (1994), and Andrea Lemon and Sarah Cathcart's *Tiger Country* (1995), all of which are at least partly set in the desert. While these plays have attracted critical attention for their use of bodies and narrative, I want to read them in terms of the poetics of space that is increasingly significant in contemporary Australian theatre. In highlighting setting as much as corporeality, language or action, these plays endeavour to map a space for women in Australia, literally and metaphorically. In doing so, they also highlight more generally the need to read subjectivity in terms of spatial context.

Spatial practice in Australia has, in Edward Said's terms cited above, been 'complex and interesting' for the two centuries of white Australia's settlement, achieved, in effect, by dispossessing indigenous peoples of their land. In the most recent expression of the ongoing anxiety regarding Australia's settlement, or 'unsettlement', as it were, the government has even altered the nation's borders to exclude certain islands from counting as 'Australian' territory, in an attempt to halt refugee claims being made there. 'Imaginings' of Australian spatiality and its resonance within the nation's 'political unconscious' thus necessarily incorporate diaspora, exile and dispossession, matters generally associated with an absence of space. Desert settings also foreground compromised space: the plays discussed here disrupt 'known' space to privilege what we could call a 'psychic' space, associated with Fensham's conception of the 'feminine unconscious'.

Space is, of course, gendered, like other socially constructed institutions; for Doreen Massey, 'this gendering of space and place both reflects *and has effects back on* the ways in which gender is constructed and understood.'[6] Landscape itself has frequently been figured as female, as Annette Kolodny describes in the American context and Kay Schaffer in Australia. In fact, Sue Best maintains that 'in an extraordinary array of contexts, space is conceived as a woman'.[7] Gillian Rose pursues this masculinist perspective which is, for her, characteristic of the study of geography generally: '[t]he geographical imagination thinks space can always be known and mapped, and that's what its transparency, its innocence, signifies: that it's infinitely knowable; that there are no obscure corners into which geographical vision cannot penetrate.'[8] The plays considered here attempt to compromise this 'knowability' by focusing on a 'known' location at the same time as they shift the setting to a psychic location, one that, in its unmapped state, appears to be 'unknown'. The women journey to/through the interior, which is both geographic and a representation of subjectivity. I explore this topographical and performative layering of locations, the spatial 'geology' of these plays. Of course, theatre inevitably operates on the construction of various imaginary locations, some of which represent real places and some of which do not, but all of which combine to form a production's diegesis. The difference in the spatial geology constructed in these plays is that the desert landscape only comes to make sense in terms of setting and subjectivity when the layers of its various manifestations are accrued, combined and contrasted.

Still Angela

> [T]here are two landscapes, Angela. One always on top of the other.[9]

Jenny Kemp argues that her work 'attempts a dialogue with the disjunction that exists when you walk down the street and see the real world but feel aware of an "inner world"'.[10] She locates on stage the 'inner landscapes of the female psyche' (Fensham, 'Modernity', p.9), exploring the lives of contemporary women through doubled characters, flashbacks and fractured narratives. Fiona Scott-Norman writes that a Kemp play:

> is about the fracturing of a single moment, and the exploration of every shard of possibility, every perspective, every fantasy, every memory and resonance connected with that moment... [A]lthough there are only shards of meaning present, they are of sensations and situations relevant to everyone.[11]

Kemp spatializes and narrativizes each shard as a *location* of the character's subjectivity, with the eventually reconstructed shattered object resonating more productively than the original. Spatial geology is described most simply in *Still Angela* by the epigraph above: the real landscape exists with an internal, psychic space (although the presence of just two landscapes is arguable).

Still Angela traces the life of Angela, about to turn 40, as she looks back at herself through various ages: as a 6-year-old when her mother died, in her late twenties, and early thirties, each played by a different actor. At the end of the prologue, Angela 3 says: 'There was something to discover about time, it was as if the sandwich, the bushes, the trees, the earth, were all getting on with something and she just wasn't quite getting it. Something important was eluding her' (Kemp, *Still Angela*, p.2), something she comes to find by exploring her physical surroundings and her psychic landscape, and the interrelationship between the two.

In order to understand this relationship between physical and psychic, one must apprehend the nature of the physical set. The play's geographical settings are framed by her kitchen and by the desert. It is, however, difficult to determine which are real and which are not, since Kemp uses as the basis for her work the paintings of Paul Delvaux which 'remind me that the world *on stage* is a timeless place, a kind of landscape of the psyche where anything can happen'.[12] The desert, for instance, is both a literal place (the play opens with a shot of the Simpson Desert on the narrow backdrop of video screen) and an interior landscape (a glimpse of the past). Thus few locations on stage come to have just one meaning: beyond the narrow corridors of light and the barest representation of a kitchen are dimly lit (in blue) tree branches that represent an undefined 'outside'. The most articulated versions of the 'outside' come from either the video footage of the desert or the garden path that her father literally builds on stage. Yet we know that this action also takes place in her mind's eye and is one of the actions that takes Angela back to her mother's death. When Angela sits in her kitchen – a barely realized space that also comes to represent the kitchen of her childhood – she comments: 'There's a picture on the wall of the desert, it looks bare and austere, it looks as if it goes on forever' (Kemp, *Still Angela*, p.13). That psychic landscape is then animated on the video screen, which becomes the image Angela sees from the train, moving in the desert. It is left unclear, deliberately, whether or not Angela actually does travel to the desert, even though the play records the train trip she takes. Either way, psychic space comes to be more extensive than conventional space, as it spills over walls and borders to become all encompassing. Fully realized space lies over the top of psychic space for the first part of the action; once Angela visits the desert, the layering of spaces appears to be reversed. The psychic space has a transformative effect on the character of Angela: after Kemp's characters inhabit psychic space, they are subsequently more able to negotiate the fully realized and mapped physical space.

Angela's reintegration following her visit to psychic space is theatrically complex. Kemp characteristically works closely with a choreographer (Helen Herbertson) to establish a pattern of movement that is appropriate to the action: spatiality is written into the printed text in Kemp's work as both inner and outer landscapes appear to dissolve into each other. As Angela sifts through her mind and memories to try to discover what's been eluding her, the Angelas traverse the floor in an apparently random, confused pattern. '*As she sits (or at times moves) she travels internally, her memories and emotions bounce, jostle and nudge her for attention. The action should be rhythmic, as if Angela inhabits areas as she gives them focus*' (Kemp, *Still Angela*, p.3). At other times the movement is regular as the metaphor of the chess game that is introduced into the action by

Jack. John Mangan explains that the production of *Still Angela* is structured using 'corridors, layers,...and grids'.[13] These corridors and grids are provided by the chessboard, and the rules of direction for chess pieces also contribute to the movement, a chessboard itself being a representation of both real and imaginary space. The chessboard is frequently the only prop lit in this relatively dark play, and the board's pattern is also projected on to the garden path that the father builds on the stage. Angela is encouraged by her mother's voice to use her 'horse' (Kemp, *Still Angela*, p.26); the movement pattern of the knight chess piece is also reflected in the shape of the father's path. Rather than charting these various movements and locations, all of which contribute to spatializing the 'lost in thought' part of Angela, *Still Angela* suggests that a definitive mapping is unnecessary. The space appears to be alternatively random or governed by a different type of spatial logic that is not cartographic but still essential to Angela's course of direction. The layering of the psychic and 'real' landscapes is a sufficient map for the nature of Angela's current journey.

Whether or not Angela actually travels to the desert, she hears a voice-over on the train that explains the topographical transformational powers of the Simpson Desert: the otherwise dry bed of Lake Eyre can fill up with water several times a century, and the desert's former life as a radically different landscape (as ocean) is brought home to Angela when she finds a seashell on her desert walk. The topographical transformation becomes allied with the psychic, as the two landscapes continue to be layered over the top of each other. The stage directions suggest Angela '*is orientating, coming to terms with the enormity of the space around her, with fear of the dark and the unknown*' (Kemp, *Still Angela*, p.18). Following the trauma of re-enacting her mother's death, the earlier scenes are played out again, in the desert, as it were, with domestic space layered over desert space. This time,

> the form [of the action] is changed totally. This is really as a result of new energy and clarity found in the desert. Now there is some space, distance between things, and an ability to feel things physically. There is more of a sense of Angela being 'embodied'.
>
> (Kemp, *Still Angela*, p.21)

Identifying the different spaces – both literal and psychic – enables her to achieve this embodiment. Kemp's women must exist in conventional social space but they also walk through and past such literal spaces to occupy other types of space, even occupying several different types of space at the same time.

In the play's final scene, the train (itself a complicated image of gendered, psychic and psychological space) returns the Angelas to the city. In 'Back Home with the Desert Inside', Angela has integrated the experience of the desert with her life in the here and now. Her fortieth-birthday party (a masquerade party for which she dresses as a tumbleweed, still traversing landscapes) brings together all the characters of the play, many of whom are aspects of herself. The two landscapes continue to exist, one overtopping the other, their geological layers now recognized. As is typical of Kemp's work, the degree of 'reality' of any scene or sequence is less important than the extent to which it assists the character to mark her location with/for herself.

Ningali

> Well they say you can take the girl outa the country. But you can't take the country outa the girl.[14]

Whereas *Still Angela* has several actors playing the same character, *Ningali* focuses on one actor who performs a variety of 'personas', each of which forms part of her character's subjectivity. *Ningali* layers physical and psychic spaces, much as *Still Angela* does, but whereas Angela's journey to the desert is in response to a traumatic memory, Ningali's relationship to the desert is more visceral. The significance of the psychic connection to landscape comes into play more when access to the literal geographical locations is removed.

The chorus to the first song in *Ningali*, by Josie Ningali Lawford, Angela Chaplin and Robyn Archer, begins with: 'Now she's somewhere under the desert sky' (Lawford, Chaplin and Archer, *Ningali*, p.2). *Ningali*, which is part autobiographical monodrama, part stand-up comedy routine, part a performance of contemporary and indigenous music and dance, and part indigenous history, traces 'a journey across vast distance and many landscapes. It is about balancing and mediating worlds and cultures.'[15] Under a broadly configured desert sky, the play mixes indigenous form and content with Western styles as it traverses a wide geographical and geological span to compare all the possible facets of Ningali's complex, layered subjectivity, which crosses indigenous and Western cultures. The play follows Ningali from Fitzroy Crossing in northern Western Australia to Perth for high school, to Anchorage, Alaska, on an American Field Scholarship, to Sydney to train with the

Aboriginal and Torres Strait Islander Dance Theatre, and to performance in the national tour of the first Aboriginal musical, *Bran Nue Dae*. Yet it also fixes location much more specifically than may be first apparent, especially in the context of literal dispossession from one's land. Among its numerous achievements, *Ningali* argues for the significance of ancient and recent Aboriginal history and law, both of which are figured in spatial terms. More than just a personal journey, *Ningali* also documents the importance of desert landscapes to Aboriginal women (and men), which function substantially differently than they do for settler women in Australia.

If the desert's endless expanse of space represented fear, the unknown and a return to the repressed for Angela, it is for Ningali a cause of joy and pride:

> But it felt endless, there were no fences,
> We could go wherever we wanted.
> I knew all that big country of my people
> Our dreaming was all over that land.
>
> (Lawford, Chaplin and Archer, *Ningali*, p.4)

The desert (in this case, remote cattle country) is Ningali's home (as opposed to the urban Angela who seeks an understanding of herself by visiting the Simpson Desert), and the basis for her Aboriginal heritage, the foundation stories that she lives by, and the traditions which determine her culture. Access to the land is, however, taken away from Ningali and her family in 1977. Fences now separate Ningali and her family from topographical sites that had been sacred for centuries, and from more personal locations such as the tree under which Ningali was born. The severing of access to the land threatens the spiritual connection, and, in turn, the culture itself: her family asks, 'How are we going to do law properly if we can't tread on our own land?' (Lawford, Chaplin and Archer, *Ningali*, p.9).

Spatiality is figured in the play in three principal ways, each of which contributes to a type of reclamation of that lost land. In the first instance, Ningali marks the connection between dance and landscape. This is a somewhat abstract connection for many non-Aboriginal audience members, but Ningali explains dancing as a way of literally marking the land in the sand while also representing its custodianship. Dance is a moved representation of sacred and literal spatiality, as the literal and figurative use of landscape continually intertwine. Ningali recounts the history of her people – 'Maybe the oldest culture in the world' (Lawford, Chaplin and Archer, *Ningali*, p.26) – by dancing her tribal dance which transforms the space of the stage from 'just' an indoor, contemporary theatre to also encompass thousands of square kilometres of Ningali's people's land, over thousands of years.

Second, landscape intersects with corporeality as the space of the set reflects Ningali's literal body. While Ningali's narrative unfolds on the stage, the audience becomes aware of the play taking 'place' on a floor-cloth that depicts a somewhat abstracted image of the performer herself. In other words, 'during the course of the performance, Ningali Lawford inscribes her story on the landscape of her face' (Grehan, *Mapping Cultural Identity*, p.76). The performer's body is staged spatially as well as narratively: a specific literal theatre space and a psychic space of subjectivity appear, combining individual subjectivity with a well-defined historical context, rendered spatially. At the end of the play, Ningali paints up by removing her shirt and applying paint to her face, arms and bare chest in preparation for a final dance. The markings that she applies to her skin are actual, topographical locations as well as representations of spatiality, as she inscribes literal 'space' on to her 'literal' body in yet another way.

Finally, a vestige of the land is represented through language, which acts as a literal and psychic trace of the spatially defined past. *Ningali* is performed in English, Wambajarri (her native tongue), and kriol (a combination of the two). Ningali engages in some direct translation and some contextual translation: likely few audience members will understand everything, which is partly the point. While English is used to communicate the bulk of the narrative, the spatial memory contained in each dance is encapsulated in the Wambajarri language. Formerly 'just' a means of marking the land and telling dreaming history, tribal dance now enacts a more significant function, since Ningali's family no longer has direct access to the land from which the dances, language and culture emerged. The power of the dance – itself implicitly spatial – is in some respects transferred to language, a signifier of the physical culture that the Wambajarri are lucky enough not to have lost with their lands. At the end of the play, the land – specifically the tree that marks the place of Ningali's birth – is not accessible. There are material limitations (barbed wire) and psychic drawbacks (no access to ancestral sacred sites) to these shifts in 'ownership' of the land. *Ningali* demonstrates the process by which spatiality is transferred to other media (including, here, dance, corporeality and language) when geographical access is no longer available. Ningali's family's exile from their ancestral lands does not sever the connection: instead, spatial connection is configured differently, in a type of psychic relationship that echoes Angela's psychic traversing of the desert. While their association with the desert differs, both women illustrate ways to read its spatiality – and its spatial effects on their subjectivity – in a layering of interpretational possibilities, few of which are strictly geographic.

Both *Still Angela* and *Ningali* concentrate on a relatively integrated subjectivity of one woman, whereas the second pair of plays outlines multiple subject positions in the staging of several women who combine to integrate with(in) the nature/nation of Australia.

Tiger Country

Maybe you'll find yourself one day.[16]

Lucy, an Aboriginal woman, explains to Barb, a white city woman who comes to live on a remote sheep station, that understanding herself is directly connected to understanding her landscape at both a literal and a deep level. This is no mean feat, since the desert landscape in *Tiger Country* is unusual enough to be compared to the moon (Lemon and Cathcart, *Tiger Country*, p.8) and, later, to Mars; it is the location of UFO sightings and one character even claims to have been abducted from the desert by aliens. While *Ningali* traces Aboriginal responses to the desert, *Tiger Country* charts the diverse experiences of four unrelated white women living in the remote landscape between Maralinga and Woomera, South Australia, at various points in the last 150 years. Andrea Lemon and Sarah Cathcart have produced three one-woman plays drawn from extensive interviews with diverse women. *Tiger Country*, following *The Serpent's Fall* (1988) and *Walking on Sticks* (1991), is the first to concentrate on a particular geographical location, rather than a specific situation or theme. In *Tiger Country*, Cathcart embodies all four characters as they each negotiate their ways literally and figuratively in landscape through time. Even though the setting is a relatively discrete geographical area, each woman lives in a vastly different landscape, yet each landscape is characterized by the 'unknown.' The programme explains that 'TIGER COUNTRY is out there . . . beyond the border . . . beyond the boundaries of our experience and understanding.'[17] *Tiger Country* records responses to a landscape that is essentially unknown to most Australians. This landscape's remoteness only partly encapsulates its unknown qualities: it is also officially 'unknowable' since this area is also associated with the contamination of the Maralinga nuclear bomb tests conducted in the 1950s and limited access to Woomera,[18] a town whose rocket test sites and satellites rendered it (then) all but prohibited to outsiders.

Tiger Country's women are Louisa, arriving in Australia 150 years ago without any idea of what to expect; eccentric Iris, aged in her seventies, an opal miner around the time of the Maralinga tests and who now runs a roadhouse; city-bred Barb, who now lives on a sheep station bordering Maralinga lands; and 12-year-old Stella who lives in Woomera in the 1970s, the height of its rocket range days. Another character, Lucy, an older indigenous woman, appears in three of the narratives to link the women: although Cathcart doesn't embody her, Lucy lives near Barb's station, stops at Iris's roadhouse, and visits Stella's secret cave. The experiences of the women layer various responses to the unknown that describes – literally and figuratively – the remote interior landscape of Australia.

The production's raked stage is bare except for a low rostrum on which is painted a large compass pointing to north, or, as the epigraph to this section suggests, possibly pointing to the idea of 'direction' itself. Occasional images (mathematical formulae, a map of the heavens, a topographical chart) are projected as a backdrop, as each character outlines her relationship with the strangely mutable landscape that was, as two characters note, once a sea. The women's narratives are interwoven to echo, contradict and occasionally complement each other in their attempt to approach the landscape that continues to be associated with the unknown.

The women do not all succeed in negotiating the unknown: in particular, Louisa experiences terror in the indescribability of Australian space: 'I could find this countryside almost charming... if only there were something here' (Lemon and Cathcart, *Tiger Country*, p.36). The remainder of this section looks at the ways the other three characters read a landscape that Louisa can only see as empty. Barb begins the same way as Louisa: 'All this red sand is going to kill me. It starts at the verandah and just keeps going... out to the horizon. My backyard is 2000 square miles... I feel like I'm drowning' (Lemon and Cathcart, *Tiger Country*, p.31). Barb comes to read the space around her differently, though, when she meets Lucy who takes her to a corroboree. Lucy's engagement with the landscape, including her perception of the beginning of the world (Lemon and Cathcart, *Tiger Country*, p.39), helps Barb appreciate the different ways in which she might understand and interpret the land. Barb comes to recognize the landscape by being able to understand its multiple meanings for all those who inhabit it: ironically, it becomes known to her when she understands that it means so much to many other people. Barb's lesson provides another example of the geological layering of various locations, all of which contribute to a more complete map of these women's desert.

Iris and Stella, who respond to the landscape and to the unknown infinitely better than the other two women, expand the definitions of 'location', all of which continue to accrue in this exploration of the known and unknown desert. Helena Grehan argues that the play's landscape 'is more than a view or stretch of land; it is a surface which encompasses sky, below and above ground' (Grehan, *Mapping Cultural Identity*, p.49). While Barb understands the landscape by reading it in conjunction with Lucy's history, Iris digs into it – literally – to mine opals. Iris remains one of the most practical of the characters in her understanding life in the desert: she checks her roadhouse customers to ensure their cars are in sufficient condition to travel the isolated distances. She insists that they have enough water, declaring: 'Y'gotta find out about the country you're goin into mate... y'gotta understand it... y'gotta treat it with respect!' (Lemon and Cathcart, *Tiger Country*, p.18). Yet it is not sufficient to label Iris as 'just' practical because Iris herself is layered with another layer of subjectivity that is associated with landscape, or at least losing touch with landscape: Iris claims to have been abducted by aliens in the lead up to the Maralinga tests, and her description of the events emphasizes that she accepts both her practicality and her alien experience on the same spatial plane. Iris's two perspectives on the desert (the practicality of water and petrol, balanced with the abduction story) exemplify what each woman demonstrates: the layering of a different configuration of geographical landscape with a less mappable perception of literal and figurative spatiality.

Stella talks about the function of her town, the satellite tracking station, from the perspective of 'knowing' it all: 'we know what's going on everywhere in the world... everywhere in the universe' (Lemon and Cathcart, *Tiger Country*, p.21). This knowledge enables Stella to frame her surroundings by means of a rational, scientific model. Yet in her world of top-level surveillance, Stella learns the significance of a more private, psychic connection with the landscape that is essential for survival. When a missile crisis results in the Americans being evacuated from the country, Stella's fascination with space shifts from a scientific, all-knowing perception of the heavens to her special earthly place, a cave with Aboriginal paintings. When Lucy appears at 'her' cave, Stella comes to understand that the scientific model is only one possibility for interpreting her landscape, and that Lucy's approach might also be appropriate.

The women's experiences overlap even though it is only Lucy, the Aboriginal woman who does not appear on stage, who directly crosses over the narratives. Yet the women traverse similar space, and the experiences of negotiating the unknown entwine them in comparable narratives as they occupy the same desert space. Of course the women also occupy the same stage space – even the space of the same actor who plays them all. Towards the end of the play, in a scene called 'Parallel Universe', all four women briefly discuss much the same issue: the magnitude of the universe. Stella completes the scene, declaring in her love of outer space matters, 'they reckon spaceships might be windows into one of these parallel universes!' (Lemon and Cathcart, *Tiger Country*, p.44). The scene constructs the women as parallel to each other, but it also reinforces the different ways in which they read a paralleling of 'real' desert with a balancing, psychic view of that landscape.

This scene is the core of the play: each of the women offers possibilities for reading and negotiating the landscape, but no possibility is completely accepted or discounted. Rather, the play fosters a layering of all the possible interpretations to help reconfigure the nature of landscape, culture and self. This manoeuvre is significant in two key respects: first, the investigation of their various landscapes becomes immediately personal for these women since *Tiger Country* provides four significantly different and individual interpretations of the landscape, the unknown and the self. A surveyor's accuracy is not at issue: instead, as the epigraph to this section indicates, the women come to find appropriate, personal maps of the self which in turn provide a way to live in the land that surrounds them. The unknown quality of the space beyond the borders comes to be less amorphous for at least three of the women. Second, the play reinforces the incompleteness of what Gillian Rose, cited above, terms masculinist geography wherein 'exploration' (and conventional exploration narrative could also be included) sets out to map – indeed, capture – every detail of topography. As *Tiger Country* demonstrates, the revelations about location that are most meaningful to the characters are not necessarily able to be fixed as geographic 'locations'. By dramatizing both the literal unknown (for instance, land contaminated by nuclear tests and therefore prohibited territory to most Australians) and the psychic unknown (such as the characters' traversing of parallel universes), in addition to other types of possible locations (Iris's alien abduction), *Tiger Country* reinforces the potential for a reconfiguration of a diverse, layered mapping of subjectivity in geography and beyond. *Tiger Country* provides various opportunities for the characters to 'find themselves' – literally and figuratively. While Iris aims for the underground, both she and Stella also find some solace in outer space, whereas the final play takes the psychic into a different location altogether: cyberspace.

Historia

> Here in this Australian desert with you
> I have become the part of yourself
> You hide in a dark exile of Polish words
> Far away from the starched, cricket-white
> Of your English.[19]

Noëlle Janaczewska's work characteristically combines several cultures under a broad umbrella of 'Australia'. She incorporates many languages and perspectives to create dialogue between the official approach to 'Australia' (increasingly Anglo-Celtic with the current conservative political climate) and the reality of the cultural mix (increasingly multi-cultural and multi-lingual). *Historia* configures cyberspace as a possible psychic space that is layered with geographical space to facilitate another type of exploration of real/virtual. Yet rather than seeking an integrated subject, *Historia*, like *Tiger Country*, demonstrates a broad spectrum of spatial alternatives in/for Australia. While no one person can inhabit them all, the nation's inhabitants need to be aware of their existence in order to begin to understand the country, its landscape and its space.

Historia traces the paths of two women, Zosia and Zoe, who are having an affair. While Zoe tries to keep the relationship from her husband, Zosia is trying to both reconcile her Polish background and reconstruct the story of the Polish ethnographer Bronislaw Malinowski who arrived in Australia in 1914, and his lover, a Polish artist named Stanislaw Ignacy Witkiewicz. Malinowski is about to leave him for Elsie, a Melbourne woman. Zosia is researching Malinowski, deciding that the 'history' that she has uncovered in English differs from the Polish: 'The two languages tell different stories' (Janaczewska, *Historia*, p.273). As is typical in Janaczewska's work, language often sets the scene as word lists from the two languages are recited and even projected onto a backdrop to bring several cultures together, with inevitably semi-compatible – if poetic – results.

In addition to this linguistic 'setting', *Historia* is geographically set partly in the desert, and partly in Penrith, outer-suburban western Sydney. The windy, cultural desolation of Penrith is compared to the isolation of the desert that the anthropologist explores. Throughout the play, the wind howls and pages of books rustle and flutter as time and space are combined visually and evocatively. The literal space of the desert vies with the stifling family values of Penrith, connected by a third space, a psychic space that is made visible through cyberspace as Zosia and Zoe communicate by email and as Zosia tries to uncover more research via her computer modem. Language appears to be the dominant element in *Historia*, but it is only through cyberspace that language is actually 'located'. In other words, language is literally staged 'as' space by being captured on a computer screen and projected on to the theatre's wall. Language comes to be seen as a 'place' that can be visited by means of the larger location of cyberspace. *Historia* spatializes language, compares the locations of Penrith and the desert, and actualizes cyberspace as an equivalent place.

At the beginning of the play, Malinowski and Witkiewicz appear to be conjured up on stage by Zosia's computer. Zosia retraces the steps of the expedition 'into the interior' (Janaczewska, *Historia*, p.261), although she does it from her desk in Penrith, as opposed to 'literally' visiting the desert. She says, 'I know they came this way' (Janaczewska, *Historia*, p.261), referring to the expedition and to the relationship between the men which mirrors her relationship with Zoe: 'I suspect there's more to that desert/ expedition than meets the eye' (Janaczewska, *Historia*, p.262). Her attempts to break the coded messages have a dual purpose: they contribute to historical knowledge, and they explore the complications of her own Polish heritage, or more specifically, how one manages one's semblance of Polish heritage in Australia. She tries to 'Seek another version/ Of The Artist and The Anthropologist' (Janaczewska, *Historia*, p.272), as she seeks another space for herself. The two Australian-Polish histories are superimposed on each other, each giving shape to the other. But in supporting each other, these histories develop to gather enough detail to reveal the cracks in the story: Zosia finds evidence of Malinowski trying to paper over any cracks in his history as he marries Elsie, and abandons Witkiewicz (Janaczewska, *Historia*, p.281).

Email makes the words and world visible, rendered in space. The geography of the anthropologist's journey is thus shifted into a different spatial dimension that can be enacted theatrically. The language of email, and its topic, love, are both specifically geographic, as Zosia relates to Zoe, appropriately by an email message:

> Your words fall down like rain
> Upon a violet sea.
> Your message superimposed over mine

Spreads your body across
The matrix of the screen,
Across the horizontality of the land;
And I no longer try
To understand the language,
But experience instead the fiery shimmer of colour
That polishes laughter.

(Janaczewska, *Historia*, p.284)

The two pairs of lovers talk about love extending beyond the limits of geography. More importantly, the superimposition of words and bodies, made possible by cyberspace, provides the key to Zosia's expedition through history and her personal space. The glow of the computer screen connects to the glow of the stars of the desert or the neon of suburban shops and the lights of passing trains, as the various worlds are brought into relation with each other. Computer-based '[t]echnologies... often serve to alter bodily states, thus constructing alternative bodily spaces, activities and experiences',[20] but in this case they also construct an alternative space altogether, one that nevertheless intersects with the space of history and more conventional geography.

Historia enacts a geological layering of the versions of Malinowski's story, Zosia's story, Zoe's, and the combinations therein. Zosia characterizes these by detailing the cracks in the plaster, which successive layers of wallpaper try unsuccessfully to hide. The flutter of the pages of history books reinforces the point. In the unsuccessful covering up of the past, Zosia is able to discover what she needs to do: 'I need to find other patterns in my cultural background. [*Pause*] I'm looking for a way to be Polish-Australian that doesn't gloss over, or make invisible, my sexuality' (Janaczewska, *Historia*, p.283). She then decides regarding Malinowski and Witkiewicz: 'I have uncovered different options;/ Other histories forged in Polish letters.... And their expression/ Gives me a place to be myself' (Janaczewska, *Historia*, p.284). The superimposition provides the foundation for her own, multi-layered literal and psychic location. Its spatialization on stage permits her to think about mapping it on to her self. Zosia and Zoe do not travel literally to the desert, but Zosia's cyber-trip to the desert to discover 'another version' of Malinowski's history provides her with a way to read her own life story. The establishment of a psychic space enables her to read Malinowski in the languages of English and Polish, and even the language of 'love'. From this interpretation of languages and spaces in which such languages are comprehensible, she decides not to submit to the exile of which the Artist speaks in the epigraph to this section. The space of the desert combines with cyberspace, offering Zosia the possibilities of mapping herself in personal, professional and cultural terms.

These plays illustrate numerous ways in which women can occupy the space of the desert (however real or metaphoric) in multiple dimensions at the same time. In each case, the desert is counter-pointed to another psychic space, whether domestic, urban or cyber. Once the women in these plays inhabit the more intimate but paradoxically more extensive psychic space, they are subsequently more able to negotiate the physical spaces that they must traverse.

In these plays, the exploration of space (physical and psychic) comes to be of central concern. In the context of spatial theory, such a focus on location is not surprising. For instance, Henri Lefebvre, perhaps the best-known commentator on spatial studies, maintains that the presence of space precedes 'the appearance in it of [social] actors'.[21] In other words, he suggests that cultures are generated by space, as opposed to spaces being generated following cultural 'formation'. It is, for Lefebvre, an oversight to disregard space and its 'prior' existence in examining both cultural and individual subjectivity. The mapping project in these plays highlights the need to understand subjectivity in terms of spatial context by illustrating the various and different ways in which particular spaces inform subjectivity for most of the characters.

Of course, this 'mapping' project is more specifically gendered than Lefebvre's and in these terms, the choice of the desert as a means to explore this thematic is of special relevance to the question of feminist theatre futures. In navigating possible ways to read what could be considered 'abstract' spaces, these plays specifically chart space(s) for women that extend the boundaries of map well beyond conventional geography. Spaces that fail to register on conventional maps and which can embrace the real and metaphoric, geographic and psychic, the past and the present, continue to be important in imagining and producing future 'places' of/for women in theatre and in wider cultural contexts,

Further, in a global framework, in the wake of the increasing populations of people around the world – both female and male – whose claims to a literal, geographical place have been compromised by contemporary geopolitics, issues relating to the staging and interpretation of psychic locations are becoming increasingly vital. A consideration of the sorts of questions opened up by these plays should therefore play a key role in thinking around a feminist future, in both political and theatrical contexts.

Notes

1 Irit Rogoff, *Terra Infirma: Geography's Visual Culture* (London and New York: Routledge, 2000), p.28.
2 Kevin Hetherington, *Expressions of Identity: Space, Performance, Politics* (London: Sage, 1998), p.105.
3 Edward Said, *Culture and Imperialism* (New York: Knopf, 1993), p.7.
4 For a thorough exploration of the use of desert in literature, art and film, see Roslynn Haynes, *Seeking the Centre: The Australian Desert in Literature, Art and Film* (Sydney: Cambridge University Press, 1998).
5 Rachel Fensham, 'Modernity and the white imaginary in Australian feminist theatre', *Hecate*, 29 (2003): 7–18, p.16. Fensham's perception of a feminine unconscious draws on Rita Felski's *The Gender of Modernity*, (Cambridge, MA, and London: Harvard University Press, 1995) and is exemplified in the work of the Australian playwright and director, Jenny Kemp. Fensham's reading of Kemp's work incorporates the political unconscious of Jennifer Rutherford's *The Gauche Intruder: Freud, Lacan, and the White Australian Fantasy* (Melbourne: Melbourne University Press, 2000), which focuses on Australia's 'white imaginary' and its concentration in the metaphoric and literal space of the desert.
6 Doreen Massey, *Space, Place and Gender* (Cambridge: Polity, 1994), p.186 (emphasis in original).
7 Sue Best, 'Sexualizing Space', in Elizabeth Grosz and Elspeth Probyn, eds, *Sexy Bodies: The Strange Carnalities of Feminism* (London and New York: Routledge, 1995), pp.181–94, p.181. The gendered nature of spatiality reveals contradictions that Best attempts to elucidate:

> feminising space seems to suggest, on the one hand, the production of a safe, familiar, clearly defined entity, which, because it is female, should be appropriately docile or able to be dominated. But, on the other hand, this very same production also underscores an anxiety about this 'entity' and the precariousness of its boundedness.
>
> (Best, 'Sexualizing Space', p.183)

Thus, many feminist critics note that the space of 'home' is also associated with women: see, for instance, Massey, *Space, Place and Gender*, p.180.
8 Gillian Rose, 'Some Notes Towards Thinking about the Spaces of the Future', in Jon Bird, Barry Curtis, Tim Putnam, George Robertson, Lisa Tickner, eds, *Mapping the Futures: Local Cultures, Global Change* (London: Routledge, 1993), pp.70–83, p.70.
9 Jenny Kemp, *Still Angela* (Sydney: Currency, 2002), p.19.
10 Rachel Fensham, 'Making a mythopoetic theatre: Jenny Kemp as director of an imaginary future-past-present', *Australasian Drama Studies*, 44 (2004): 52–64, p.52.
11 Fiona Scott-Norman, 'Review of *Black Sequin Dress* by Jenny Kemp', *Bulletin*, 9 (1996). Reprinted in *ANZTR: Australian & New Zealand Theatre Record*, March 1996: 52.
12 Jenny Kemp, 'A Dialogue with Disjunction' in Virginia Baxter (ed.), *Telling Time: Celebrating Ten Years of Women Writing for Performance* (Sydney: Playworks, 1996), pp.28–31, p.30.

13 John Mangan, 'Many faces, and all of them Angela', *The Age*, 9 April 2002: Culture 4.
14 Josie Ningali Lawford, Angela Chaplin, Robyn Archer. *Ningali: The Story So Far* (unpublished ms, 1994), p.23.
15 Helena Grehan, *Mapping Cultural Identity in Contemporary Australian Performance* (Brussels: Peter Lang, 2001), p.75.
16 Lucy, the unseen character, speaks these lines to Barb in Andrea Lemon and Sarah Cathcart, *Tiger Country* (unpublished ms, 1996), p.57.
17 Andrea Lemon and Sarah Cathcart, programme notes to *Tiger Country*, n.p.
18 Since then, Woomera was rendered prohibited in another respect when it became the best-known detention centre for asylum seekers. The centre has since been closed. While this figures in audience associations with Woomera now, it would not have been relevant when the play was written.
19 The Artist speaks these lines to the Anthropologist in Noëlle Janaczewska. *Historia*, in Peta Tait and Elizabeth Schafer, eds, *Australian Women's Drama: Texts and Feminisms* (Sydney: Currency, 1997), pp.254–84, p.283.
20 Nicola Green, 'Strange yet Stylish Headgear: Virtual Reality Consumption and the Construction of Gender', in Eileen Green and Alison Adam, eds, *Virtual Gender: Technology, Consumption and Identity* (London: Routledge, 2001), pp.150–72, p.158.
21 Henri Lefebvre, *The Production of Space* (Oxford: Blackwell, 1991), p.57.

13

Angry Again? – New York Women Artists and Feminist Futures

Lenora Champagne, Clarinda Mac Low, Ruth Margraff and Fiona Templeton

In January 2005 we met with four women New York artists to talk 'feminist futures'. It was a cross-generational meeting involving Lenora Champagne and Fiona Templeton who were part of the 1980s explosion of women performance artists in North America; Clarinda Mac Low and Ruth Margraff from younger generations of artists. Also underpinning the exchange was the way in which the artists come from different performance backgrounds and have different styles of work – though if they have one thing in common it is the way in which work by all of these women is hard to 'categorize'. With a background in experimental dance, Clarinda describes her work as 'conceptual performance art' in which the 'concept is almost more important than the execution'. She works on politically motivated concepts such as what she argues as the contemporary obsession with consuming and discarding. Influenced by the Situationists, she collaborates with others on intervening into everyday life: creating double-take situations in which people suddenly find themselves part of a theatrical intervention (like being picked up and carried across the street!). As a playwright Ruth experiments with language and creates text for theatre in which music is always present: integral to her experimentation is a playing with the traditions of the opera, or the musical, in a way that challenges the more conservative (well-made play) traditions of writing for theatre. Lenora and Fiona both have long careers as performance artists in which the women's solo has been a feature. Both have solo scripts in Out From Under, *the seminal collection of texts by women performance artists that Lenora edited in 1990.*[1] *It was a collection that brought together 'just a few of the [then] female artists', with their 'gritty gutsiness', 'fundamental passions', their anger and rage that was inspirational for feminists both sides of the Atlantic (*Out From Under, *p.xiii). Our meeting in 2005 created the opportunity to look back at that moment of anger in a feminist* *performance history and forward to the contemporary moment: to look at the changes to feminist politics, aesthetics, and artistic modes of process and production in a renewed moment of transatlantic exchange.*

(Editors)

Gerry: *Lenora and Fiona, you were part of that immensely explosive period in New York when there were hundreds of women artists who began to make work . . .*

Lenora: And I'm still working! *(laughter)*

Gerry: We felt that explosion across the Atlantic. If we were having this conversation 20 years ago or even ten years ago, we would have a very strong sense of New York women artists. But somehow that two-way communication, which I do think was partly, if not largely, to do with the women's/gay/lesbian network seems to have changed. There is some exchange – Lois Weaver works across in the United Kingdom, and Annie Sprinkle is regularly over, and Fiona, you still work both sides of the Atlantic, but it feels more tenuous than before.

Fiona: It used to be easy, going back and forth between the United Kingdom and New York. But now when I leave here [NewYork] nobody knows what I'm doing. People over there still pay some attention to what's happening here, probably because there's more media about it and it's more mainstream, but people here have absolutely no idea what's happening over there!

Gerry: So is there a falling off of women's work? Has it lost the visibility it once had?

Ruth: You probably know about the NYSCA [New York State Council for the Arts] survey that was done, I think two or three years ago. It's online and you can download it.[2] I can't remember the exact figures, but the number of women playwrights getting produced on regional theatre stages was incredibly small. I think I've blocked the percentages out of my head. It was so depressing. And of new plays it was even smaller. Everyone thinks of Paula Vogel whose plays are being done everywhere. But it's the same plays out there by one or two female playwrights. It's a very small percentage of women who are actually being produced. And new plays get shoved further down than that.

Lenora: There's been a move away from an interest in feminism in the publishing world too. There's not a lot of interest in putting out a book of plays by women right now, compared to 10/15 years ago say, when I did my collection in 1990.

Fiona: I actually think it's probably unlikely that there isn't the interest. I think the difference is that the publishers don't think there's the interest.

Lenora: Well publishers argue that plays don't sell.

Fiona: It's kind of very tacit though: the Theatre Communications Group playtext catalogue has loads of new writing from Britain, loads of new writing from Canada and almost no – apart from the obvious ones – American new writing. It just looks really strange when you see it – this big country, and America can only sell foreign plays.

Elaine: I know that many of the American feminist theatre scholars are heavily into studying the British scene which also seems to be indicative of something.

Ruth: You have such great writers. Caryl Churchill can't go wrong. Everything she does is just golden, and the late Sarah Kane . . .

Elaine: Kane is someone that has suddenly started to really figure in American scholarship, following on from the European interest in her work.

Lenora: It's also the case that if you can get work out and reviewed in London, or at the Edinburgh festival say, then it might lead to major productions in America. Of course the whole reviewing scene is another issue. Rachel Dickstein's piece on Edith Wharton is getting lots of reviews,[3] but she has done a lot of work with a new playwright Barbara Weichmann, which didn't get that kind of attention. And the same goes for new dance performances. People who do classics get reviewed. So I think that for new work, there's a real issue with getting it done in America. I think if you are doing new work, you have a better chance of getting your work on in Britain than over here.

Ruth: Looking at my male playwright colleagues, it always seems to me that when they do something, it is seen as exciting and groundbreaking. But if a woman does it, she's crazy and it doesn't make sense. And I think how does this other language playwright who's male make sense any more than the female playwrights who are trying to do something non-linear?

Clarinda: Because they're supposed to be non-linear. Women are supposed to be non-linear.

Lenora: And that goes back to Carolee Schneeman saying, 'I was talking to a structuralist filmmaker who said to me we think of you as a dancer' when she was a filmmaker. It's that whole thing about having your work seen with equality. That's still a challenge. Jill Johnston really championed postmodern dance.[4] She was a voice that was explaining it and articulating it and I think we don't have the critics to talk about it in a more insightful way. I think in dance you have people who can talk about new work better than in theatre.

Clarinda: I would say the complete opposite. Since Jill Johnston there hasn't really been anyone in terms of women's dance. This is even more upsetting in some ways because the percentage of women making dance and movement work is so much higher than the percentage of men and yet, at the top level, it's two women and 17 men or something like that.

Lenora: That's the same in performance.

Clarinda: Something is wrong with this picture. Maybe in dance it's that women get strange ideas about bodies! But then you think well Yvonne Rainer was important. It's extraordinary, it's not even a man's province and yet still about their ideas.

The new burlesque

Elaine: So how would you characterize the contemporary women's performance scene in New York?

Lenora: I find that in women's performance and in performance in general there is much more emphasis on craft now: on performers being really good.

Clarinda: Yes, that's true.

Lenora: I took my students to Intar to see Nilaja Sun and Carmelita Tropicana.[5] They were hard on Carmelita Tropicana, but they loved Nilaja. They loved what a great performer she was.

Elaine: We've noticed that women's burlesque has a New York presence and wondered why that has become popular?

Lenora: I think it's related partly to the success and interest in the work of Annie Sprinkle. The whole carnie idea of the burlesque has to do with younger women's attitudes towards sexuality. It's much more about display. Their idea – I don't know whether they use the word feminist or identify themselves as feminist – but they see themselves as having control or power over their bodies. It's in the pop culture that they've grown up with.

Clarinda: Also, in terms of popular culture, I think people are always reaching back to find the tools that have been dissipated. Making women central rather than peripheral is part of it too. They take centre stage and they're not tools of somebody else. They're creating this strange or slightly scary, slightly threatening, very sexy, very circus-like atmosphere.

Lenora: I think back to Liz Prince, back when she was doing her costumes and performing. That was borrowing on burlesque traditions.... And

Julie Atlas Muz did this piece: *I am the Moon and You are the Man on Me* [P.S.122, November 2004]. I didn't see it but I heard that she goes upside down and takes a flag out of her ass or out of her vagina...

Clarinda: It's really about imperialism and it's very clear when you see it. The set up is that she's naked, but all in white make-up and then there's a little astronaut with a little flag and eventually the guy ends up on her butt. The way she performs is like a moon goddess. You really feel like this is not a person, then this little flag goes in her ass and it's like there's something really wrong with that. You get a sense of overtaking the woman's body, like the land being overtaken.

To me that's not burlesque though. With burlesque it's like there's two schools – there's the 'messy burlesque' school and then there's the real 'high art', not really high art, but more thoughtful, generally more crafted and more clean and more shaped in a way that you can say, 'Ah, that's theatre'.

What the burlesque does is to bring different body types into the sexy arena, which is part of what I really like. There are a lot of really big women doing it and they're really successful. They have a skill, but it's a different kind of skill. I think the skill is in the moment-to-moment connection. It's a more immediate and visceral connection: 'I'm here with you, if something happens I have to react to it.'

Gerry: This puts me in mind of Marisa Carnesky in England. The connection is that I've heard Carnesky say – this was on a Glasgay conference panel – that Annie Sprinkle is an important role model for her work. One of her first shows was called The Jewess Tattooess [1999]. *She has an extraordinary body, she's tattooed all over, but actually it's an amazing classically Edwardian, Victorian Body.*

Lenora: Personally that's never been the form of performance that I'm particularly interested in, but I can see that it's a form that a number of people are very interested in.

Elaine: From what you're saying, is it something that younger women, today's student generation, see as important?

Clarinda: But, they're not so young.

Lenora: They are to me *(laughter)*.

The F word

Gerry: Turning to feminist futures – big question mark – can we ask about whether gender issues inform your work at all? Do you have any sort of relationship at all to the F word?

Clarinda: I think I have a huge, but oblique relationship to it. Sarah Schulman [playwright and novelist] once said to me, 'You make the most feminist performance I've ever seen', or something like that. I think it has something to do with the fact that my life has been involved with creating a lot of performance with, almost exclusively, women.

Lenora: I think my work has always been very involved with feminism and I think of my work as documenting different stages of a woman's life. Not just my life, but the life of my generation. I think of it as very much about dealing with issues of being a woman in the world and with identity as a woman and how that shifts given the period of your life that you're in. And I think that's probably more typical of my generation than it is of yours [Clarinda's]. I think it is women who are in their late forties and over who really started being introduced to feminism in the 1970s who are concerned with that kind of issue.

I don't think Feminism *per se*, as a capital F, really does have a lot of a future with young women. Young people don't even support abortion. They're in favour of the foetus because of foetal rights. They don't understand what an oppression it was to women to have to give birth to a child. They don't understand that giving a child up for adoption is not a solution for many women. For me feminism was more about developing an identity as a woman, than being oppositional.

Ruth: I was thinking about your work [Lenora's and Fiona's] and thinking about the kind of oppositional work of earlier feminism. But I think I could argue you both as composers of feminism rather than directors of feminism – because I always think composers work inside of the text and it seems like your feminism works inside of the feminism that was being pressed out there, in a way that has a grace to it, that is really beautiful. And I think sometimes this is even more effective than the onslaught of feminism.

In terms of feminism in my own work, I think there are times when you are put in a position where you can't be anything but a feminist: where you're up against a wall and you're being assaulted in some way by your culture. I've always thought those moments are Capital F Feminist in my work – like in *Judges* [*Judges 19: Black Lung Exhaling*, 2001–4] there's a monologue by the concubine where I think that she is a feminist in that moment. She is like Holly Hughes or Karen Finley. She's preaching and she's angry and she comes back at men because they're gang-raping and beating her to death. So I think in that moment she is pressed into this. I feel like it's a spectrum – that we're always feminist because we're women and we're in this culture that creates a lot of sexism and misogyny, so it's always there but sometimes it's in a small amount and sometimes it's full blown because you're in a certain position and you're called upon to play against something that is attacking you. So I think of it that way, like a spectrum.

Gerry: When I'm teaching students and we look at feminism and gender issues I try to tell them about the things that have changed in my generation – I was born in 1957. I try and explain to students about the huge changes in women's rights, but they just look at you and go 'What's she on about?'

Lenora: For a lot of them, their mothers were feminists and there you have it. It's like the whole Doris Lessing *Golden Notebook* thing of your kid rebelling against what you were.

Clarinda: I think it also has to do with the fluidity of sexuality. I think that for younger generations it's much more natural and fluid – a relationship of desire and gender that is much more fluid. And because the laws are also different, like my ex-girlfriend's mother couldn't get a bank account with her own name on it, if she left her husband in 73.

Lenora: In 73!

Clarinda: Yes, in Massachusetts.

Lenora: You think about Ibsen, like she [Nora] couldn't sign a loan thing, but in 1973!

A feminist aesthetics?

Elaine: Would you say that the pioneering efforts of women to write differently, to strive for a feminist aesthetics have made a difference – or has all of this just disappeared?

Lenora: Oh I think it has made a difference. I think it's still there in the writing. I just think it's not on the front page of the *New York Times*.

Clarinda: But at some point you have to ask how important is it to be acceptable to the mainstream, or do you create another kind of network? I've been thinking about this a lot. It's like the model no longer works and I think this has a lot to do with feminism. The people that I was inspired by, they were all feminist, they were all lesbians, they were strong, they were wild and that was because they had to do *that* and that was what pushed me forward. But that creates a different kind of atmosphere. It wasn't just girls and lesbians who were interested in them. It was the world – there was a world interested in them. But things shift and things keep shifting and there are different types of work.

Fiona: People are always asking me whether I make theatre or performance art, which I've always thought is a kind of useless distinction, because performance art for me began partly from the visual arts: as the individual extending the visual palette into the body and time. For me, the difference between performance art and theatre, then, was that performance art was about being an individual and being experimental and not necessarily being part of the theatre institution. It wasn't necessarily an essential difference, like an ontological difference. But I think that people have been making those distinctions for so long between performance art and theatre that those two strands are becoming the way that people think about performance. It either has less production value and is made by an individual who does everything and is about the individual, or is about making a show, making a big show, and getting all the edges rounded. It's interesting that in women's work it seems to have come down to those two things, but I think there's other work going on that doesn't fit into those two scenarios.

Clarinda: Also I'd say that I don't feel work is divided on male/female lines now, like it used to be. There's more cross-over.

Lenora: David Cale – his writing and the way he approaches language – is someone who to me feels like he's informed by women's work. I don't know whether people would agree with that or think that's fair.

'Making' space

Elaine: And what about the structures – creative and financial – for making work?

Clarinda: I'm interested in how we can create the structures in which to do the work we want to do – to create a network that isn't hierarchical like things are now. You can see how the set up in the eighties – I mean hierarchies – are not working out.

Fiona: The eighties were a big distraction.

Lenora: We're living in a period and this is now the fifth or sixth year where the hierarchy is the model again.

Clarinda: It's always the model though.

Lenora: It probably is, but I think there's this whole interest in masculinity now. My last show was a solo piece that was somehow referencing this return to the patriarchy. I used this booklet called *Mother's Little Helper* that my mother used to teach me about sex – it's from the fifties. And I quoted from it and I said how now that Bush and his cohorts were busy bringing us back to the fifties, I thought that I should remind myself of where we were headed. But also the other thing that was layered in with it was 9/11 and the anxiety of safety that that implies. It was very much about how you develop: how girls or women grow in this time where there is this fear, and you're being told that abstinence education is how you should learn about sexuality. I did have a rant about Bush!

Ruth: I think it's all about this global corporate culture now. You see it in the manoeuvres of Bush, for example. I think that his masculinity is so unilateral and arrogant and I see the arts behaving like that now. There are arts corporations also and if you are not incorporated by the theatre institutions you won't survive. You'd better fall in step because you'll be crushed like a bug. And I don't have that same optimism about being able to fell that, because I've tried to kick at the pillars a little bit. I am at a crisis point where I don't know how to do my work unless there is some kind of what I call 'institutional intuition'. Every resource that I need is linked to conforming to a marketing strategy for use value and I'm trying to be an independent artist in that world, because I've just graduated from New Dramatists,[6] and graduated from the Here Arts Center artist residency programme in New York. I'm going to be out there alone in terms of not having the 501(C)(3) to apply for what I need. It's really scary because I have these dreams and I feel like I've worked really hard to train myself and to train my mind through the conversations of the work and I'm ready to work, but the resources are all pinned up in these very masculine, authoritarian structures that to me are so emblematic of our government. I just see what's happening in Washington and I see it happening in the arts in the same way.

Clarinda: We are at a crisis point in terms of how to create work that communicates with the public we want to reach. I don't want to go through an institution. I just want to be out there making the work.

Lenora: I find it hard because when I do work, people stay and talk to me about it afterwards, and when the audience comes I have found the public I want to speak to, but it's again access to the institutional sponsors and opportunity to have it extended and access to the press. Because I think, more and more, although I got a lot of press attention in the eighties, I feel that the press are interested in discovering the next new thing. For middle-aged artists, particularly middle-aged women artists, it's tough and I see women visual artists – women who have won considerable prizes – who have the same problem. They still have trouble getting attention at this point in their lives. I feel like if you last as long as Meredith Monk does, then maybe you can get a retrospective or something!

Fiona: That generation were different though.

Lenora: They were pioneers.

Fiona: Not only were they pioneers, but I think that they came up as part of a community. People like Meredith Monk and Yvonne Rainer, Carolee Schneeman. I think they became visible enough partly because of the pushes they were making. Our generation actually was coming just behind that. But then the eighties happened which was when artists became successful like rock stars! And I think in terms of being able to continue a creative community, we just ran into the beginning of all that greed which left us behind. I think that people in the art world are still trying to make an art world where the artists are able to show again. Something happened to that generation at least for long enough that it became visible. They were very lucky.

Ruth: In playwriting there's also an understanding that if you develop a writer for long enough, then they'll come up with a well-made play. And this is how we get the regional off-Broadway plays that could have been written by really anyone and were written, actually, by invisible committee. So if you're an experimental playwright and you've been developed through a seven-year membership, you've been emerging and emerging and now you should write your well-made play, but you can't because you are experimental, then they just think well what are we supposed to do with you? We've given you every dramaturg!

Fiona: I think women are treated as just eternally emerging in the structure that we work in. I went to Lower Manhattan Cultural Council for a meeting and it was going really well and then someone said something about younger theatre artists and I thought, I'm 53. How much longer will I be a younger theatre artist? And I talked to NYSCA and they were saying you have to have a track record. I have a 35-year track record – how long do they want!

Lenora: Well it was horrifying to me one year when nobody in my class knew who Karen Finley and Holly Hughes were. How can you not know about the 'NEA Four', but then you think well that was 1991.[7] I forget how much time has passed. I do think – on the positive side, just because I think there is a positive side – there is work that has a take up. For example, I think that somebody like Cynthia Hopkins will get booked. I really liked her recent piece, *Accidental Nostalgia* [St Ann's Warehouse, New York, January 2005], which I thought was terrific. It was directed by D. J. Mendel, who's worked a lot with Foreman and visually it reminded me of Foreman sometimes, but it was so much Cynthia's piece and Cynthia's story: the idea of the objectified body, but being controlled by the woman. I think there's something really powerful and exhilarating about a young woman taking control of her body. Even though she's showing her body being manipulated during the performance – it's always, she's always, the subject of it. Her presence is fascinating. She's such a good writer and she's an excellent singer. The show definitely has commercial possibilities. I think that is the work that's going to go somewhere.

Fiona: It goes somewhere, but there are other places to go.

Clarinda: That's what I'm saying, that there are all these alternative routes.

Lenora: But I think there are fewer alternative routes. To me Franklin Furnace is not a space anymore: it's a cyberspace, which is great, and Martha [Wilson] is the one person who's been consistently supportive. But all these spaces that used to be downtown are not available anymore. I think there are more possibilities in Brooklyn, but I started out here, not in Brooklyn. There's nothing in the East Village in terms of places to perform.

Clarinda: But what I was talking about was actual methodology rather than places. It's how people are approaching each other to create spaces. That's what can change and has to change: that's what creates alternatives. You know that all these kids get together and they make a space in Brooklyn and then they partner with other people. We have to think of something different because of the way that real estate works and the way that the world works – we're connected to everybody now, all the time. So I always talk about the geography of ideas which is different from physical geography. You can have a border with Thailand and be in New York City, but you share a border because you share ideas and you're sharing experiment, so something...

Lenora: Like International WOW?

Clarinda: Right – they do have that, don't they?

Lenora: They do, with Thailand.

Forgetting history

Elaine: In terms of your own projects and feminist and/or broadly political themes, what do you feel it is important to bring to your work at this time?

Lenora: I'm working on a piece called *TRACES/fades*. It's using Alzheimer's as a metaphor for our national memory, how we forget history. I'm planning for it to be a piece with women in their sixties and seventies as the performers. It's still in the early developmental stages, but I'm going to use projections of various kinds of writing and words fading away. So it's going to be a different move for me away from solo work for a while. It's hard to be a performer in your fifties because you know that women are supposed to be a certain way and it begins to be a challenge because people really want to see young women perform. So I'm going in the other direction and having all older women in the performance. I'm also dealing with the whole issue of memory and what happens to it.

Gerry: One of the great tropes of the postmodern period is that we've lost the sense of history.

Fiona: This is actually related to some of the stuff in *The Medead*.[8] Not just because of the older women – it's every age. I'm really struggling in the piece with how do you talk about not forgetting history, but at the same time point out that there's more than one version of it and that I'm not necessarily offering a counter narrative? It's not necessarily Christa Wolf's Medea where she's got the feminist version from Medea's point of view. What I'm trying to say is why is that story told and then that story told and why's that a different story? And how do you deal with how different political moments make up different stories? How do you remain suspicious of all of the narratives and find your own way into them? With the prospect of Bush being in for a second term, I started to rewrite my Athens section, which is the most politically modern section, with an increased urgency and an increased clarity. There's this kind of push and pull again: there is something specific to say; but then at the same time you still don't just want to tell people *the* version. You want them to ask their questions, but you want them to know *this*.

Ruth: I'm working on a trilogy of what I call 'world folk operas' that include *Café Antarisa,* which is set in Crete in 1889 and inspired by failed rebellion, and *Wellspring: An American Opera Box For The Balkans*. I'm trying to understand the Ottoman Empire and its affect especially on Greece and the Balkans. I'm very interested in how we have almost ubiquitous (and Eurocentric) adaptations of the ancient Greek dramas and mythologies all over our American stages with little to no interest in the thousands of years of modern history of Greece especially for the 400 years Greece was occupied by the Ottomans. These sorts of blind spots in the American mind are very intriguing to me. Why we don't want to look at Greece as part of the Near East or Asia Minor or as a Balkan state next door to the former Yugoslavia? I'm finding many slants of history: the Turkish side, the Greek side, the Balkan side, and all the different factions within the Balkans. I'm very interested in presenting obscure, underclass rebels, refugees and Roma gypsies as colossal and operatic and in moving the form of opera in a more Eastern direction. Looking at the power of the rebel, even in a long legacy of failed rebellions, in relation to

Empire is helping me to look at our empire and our role in it as Americans and people that live on this turf. Because I think as long as we live here, we are complicit in some way.

A question of anger

Clarinda: I think it's interesting that we keep talking about American politics, Bush and this presidency. It's robbed us of perspective because the moment is so critical – such a feeling of emergency. It's hard for us to see outside of that... And there's such a lot of reaching into history now, why are a lot of people doing that?

Lenora: To try to get perspective... because it's so unbearable otherwise...

Clarinda: I feel like I have no voice. I was thinking about the [anti-war] demonstrations, but they haven't given us a voice like before...

Lenora: Well, really the difference between the Vietnam demonstrations and these demonstrations is that, before, an electorate that was angry and in the streets mattered to the people in power, and now it doesn't.

Clarinda: So how can our voices be heard? How can we make a difference? Because I think that it's very tempting – especially for artists – to give up right now. It's really tempting. And it's also tempting to just be angry which actually isn't effective.

Lenora: I'm not sure I agree with you about this. When I show students the *Sphinxes Without Secrets* video that was made in 1990 that has Holly Hughes, Robbie McCauley, Annie Sprinkle, Rachel Rosenthal and Diamanda Galas in it, their faces look like they were run over by a truck. Because *we* are used to confrontational art, but they haven't seen anything like that. Everything has been so tame and Christian and proper. I think they're open to more 'spiritual', quote unquote, approaches. They don't want to see work like that, but I think it gets to them.

Clarinda: It wakes something up?

Lenora: It wakes something up or it's something that they have to deal with that they don't really want to look at. But I'm not saying it's the way to go. I think, maybe you're right, confrontation isn't the way to win converts or influence people at this time. But you know, I don't know what is.

Gerry: I've got Sphinxes Without Secrets *– and it exemplifies what we liked about you American girls. You were angry, you shouted. All those women who were prepared to go mad on stage.*

Lenora: But there's very little of that now. I never really got that angry until my last piece [*Mother's Little Helper*]. I think that what is not recognized right now is the need people have to see their own opinions validated on stage – that's what's valuable about work that does tap into anger. I think people feel incredibly isolated and alienated and need to feel part of a community, because we're not you know – not even in the margins. We can create margins to a certain extent, but we're all living in this world that is increasingly absorbing the margins. So you need to have the choir preaching to you so that you know you're part of a community. That you aren't alone. That we are going to be able to withstand this time and go forward.

Notes

1 *Out From Under: Texts by Women Performance Artists*, Lenora Champagne, ed., (New York: Theatre Communications Group, 1990).
2 'Report on the Status of Women: a Limited Engagement?' Http://www.americantheaterweb.com/nysca/opening.html.
3 *The House of Mirth* from the Edith Warton novel was directed and choreographed by Rachel Dickstein, dramaturg Emily Morse, Ohio Theatre, New York, January 2005.
4 Jill Johnston, author of *Lesbian Nation* (1973) and 'Dance Journal' columnist for the Village Voice in the 1970s.
5 Intar is a Latino theatre company. Nilaja Sun wrote and performed *Blues for a Gray Sun*; Carmelita Tropicana wrote and performed *With What Ass does the Cockroach Sit?* (November–December 2004).
6 The New Dramatists organization was started by Broadway writers Robert Anderson and Michael O'Hara. It offers playwrights support for seven years in the form of free xeroxing of plays, a dramaturg to work with, casting assistance, and so on. As an organization it is inclusive of more experimental writers – like Fiona, Lenora and Ruth, who have all at some point been members – although later what they find (as discussed in the interview) is that they can't find a place for their very different kinds of work in American theatre.
7 The 'NEA Four' were Karen Finley, John Fleck, Holly Hughes and Tim Miller, all of whom, charged with 'obscenity' in their art, were denied funding by the National Endowment for the Arts. The artists brought a successful lawsuit against the NEA.
8 *The Medead* is part of a cycle of four pieces of work collectively entitled 'Realities'. For further details see Geraldine Harris, 'Foreshadowings and after blows: Fiona Templeton's *The Medead* (in progress)', *Studies in Theatre and Performance*, 23: 3 (2004): 165–77.

Bibliography

Ahmad, R. *Song for a Sanctuary*, in K. George, ed., *Six Plays by Black and Asian, Women Writers*, London: Aurora Metro Press, 1993, pp.159–86.

Ahmed, S. *Strange Encounters: Embodied Others in Post-Coloniality*, London: Routledge, 2000.

Ahmed, S. *The Cultural Politics of Emotion*, Edinburgh: Edinburgh University Press, 2004.

Ali, M. *Brick Lane*, London: Doubleday, 2003.

Althusser, L. 'Ideology and Ideological State Apparatuses (Notes Towards an Investigation)', in *Lenin and Philosophy and Other Essays*, trans. B. Brewster, New York: Monthly Review Press, 1971, pp.127–88.

Anderson, L. *Autobiography*, London: Routledge, 2001.

Angier, N. *Woman: An Intimate Geography*, London: Virago Press, 1999.

Anzaldua, G. *Borderlands/La Frontera: The New Mestiza*, San Francisco: Aunt Lute Books [1987], 2nd edn 1999.

Armistead, C. 'Women directors', *Women: A Cultural Review* 5 (1994): 185–91.

Armistead, C. 'Kingdom under siege', *Guardian*, 31 May 1995.

Armstrong, I. *The Radical Aesthetic*, Oxford: Blackwell, 2000.

Aronson, A. 'The Wooster Group's *LSD (...Just The High Points...)*', *The Drama Review: Choreography (And The Wooster Group's LSD)*, 29, (1985): 64–77.

Aston, E. *An Introduction To Feminism And Theatre*, London and New York: Routledge, 1995.

Aston, E., ed., *Feminist Theatre Voices*, Loughborough: Loughborough Theatre Texts, 1997.

Aston, E. 'Staging Our Selves', in A. Donnell and P. Polkey, eds, *Representing Lives: Women and Auto/biography*, Basingstoke: Macmillan – now Palgrave Macmillan, 2000, pp.119–28.

Aston, E. *Feminist Views on the English Stage: Women Playwrights, 1990–2000*, Cambridge: Cambridge University Press, 2003.

Auerbach, E. *Mimesis: The Representation of Reality in Western Literature*, trans. Willard R. Trask, Princeton: Princeton University Press, 1953.

Auslander, P. 'Task and vision: Willem Dafoe in LSD', *The Drama Review: Choreography (And The Wooster Group's LSD)*, 29 (1985): 96–8.

Auslander, P. *Presence And Resistance, Postmodernism And Cultural Politics In Contemporary American Performance*, Michigan: Michigan University Press, 1992.

Australian Government Department of Immigration and Multicultural Affairs Borders. Fact Sheets. http://www.immi.gov.au./fact/71borders.htm accessed 2 December 2004.

Australian Government Department of the Parliamentary Library. *Research Note*. http://www.aph.gov.au/library/pubs/rn/2003–04/04rn22.pdf accessed 2 December 2004.

Baker, R. *Sex in the Future: Ancient Urges Meet Future Technology*, Basingstoke: Macmillan – now Palgrave Macmillan, 1999.

Barry, J. 'Women, Representation, and Performance Art: Northern California', in C. E. Loeffler, ed., *Performance Anthology: Source Book of California Performance Art*, San Francisco: Contemporary Arts Press, 1980, pp.439–62.

Barthes, R. *Camera Lucida: Reflections on Photography*, trans. R. Howard, New York: Hill & Wang, 1981.

Bassnett, S. *Magdalena: International Women's Experimental Theatre*, Oxford: Berg, 1989.

Baumgardner, J. and Richards, A. *Manifesta: Young Women, Feminism, and the Future*, New York: Farrar, Strauss & Giroux, 2000.

Bell, E. 'Orchids in the Arctic: Women's Autobiographical Performances as Mentoring', in L. C. Miller, J. Taylor and M. H. Carver, eds, *Voices Made Flesh: Performing Women's Autobiography*, Wisconsin: University of Wisconsin Press, 2003, pp.301–18.

Bennett, S. *Performing Nostalgia: Shifting Shakespeare And The Contemporary Past*, London and New York: Routledge, 1996.

Best, S. 'Sexualizing Space', in E. Grosz and E. Probyn, eds, *Sexy Bodies: The Strange Carnalities of Feminism*, London and New York: Routledge, 1995, pp.181–94.

Bhatti, G. K., *Behzti (Dishonour)*, London: Oberon Books, 2004.

Bloch, E. *The Principle of Hope*, Vol. I, trans. N. Plaice, S. Plaice and P. Knight, Cambridge: MIT Press, 1995.

Bloch, E. 'Something's Missing: A Discussion between Ernst Bloch and Theodor W. Adorno on the Contradictions of Utopian Longing (1964)', in *The Utopian Function of Art and Literature: Selected Essays*, trans. J. Zipes and F. Mecklenburg, Cambridge, MA: MIT Press, 1988, pp.1–17.

Bogart, A. 'Stepping out of inertia', *The Drama Review*, 27 (1983): 26–8.

Bonney, J. *Extreme Exposure: An Anthology of Solo Performance Texts from the Twentieth Century*, New York: Theatre Communications Group, 2000.

Bordo, S. *Unbearable Weight: Feminism, Western Culture and the Body*, Berkeley and Los Angeles: University of California Press, 1995.

Boston Women's Health Collective. *Our Bodies Ourselves – A Health Book By and For Women*, London: Allen Lane, 1978.

Bourne, B., Shaw, P., Shaw, P., Weaver, L. *Belle Reprieve*, in S. E. Case, ed., *Split Britches: Lesbian Practice/Feminist Performance*, London and New York: Routledge, 1996, pp.149–83.

Bradwell, M., ed., *The Bush Theatre Book*, London: Methuen, 1997.

Brah, A. *Cartographies of Diaspora: Contesting Identities*, New York: Routledge, 1996.

Brewster, Y., ed., *Black Plays*, London: Methuen, 1987.

Brewster, Y. *The Undertaker's Daughter*, London: Black Amber Books, 2004.

Broude, N. and Garrard, M. D., eds, *The Power of Feminist Art: The American Movement of the 1970s, History and Impact*, New York: Harry N. Abrams, 1994.

Bruley, S. 'Women Awake, the Experience of Consciousness-Raising' [1976], in Feminist Anthology Collective, eds, *No Turning Back: Writings from the Women's Liberation Movement 1975–80*, London: The Women's Press, 1981, pp.60–6.

Buck, L. 'Unnatural Selection', *Stilled Lives*, Edinburgh: Portfolio Gallery Catalogue, 1996.

Bush Theatre. *Bush Theatre Plays*, London: Faber & Faber, 1996.

Butler, B. *Bodies that Matter: On the Discursive Limits of 'Sex'*, London: Routledge, 1993.

Butler, B. *Undoing Gender*, New York and London: Routledge, 2004.
Carby, H. V. 'White Woman Listen! Black Feminism and the Boundaries of Sisterhood', in Centre for Contemporary Cultural Studies, eds, *The Empire Strikes Back: Race and Racism in 70s Britain*, London: Hutchinson, 1982, pp.213–35.
Carlson, M. *The Haunted Stage: The Theatre As Memory Machine*, Ann Arbor: University of Michigan Press, 2003.
Case, S.-E. *Feminism and Theatre*, Basingstoke: Macmillan, 1988.
Cathcart, S. and Lemon, A., *The Serpent's Fall*, Sydney: Currency, 1988.
Cathcart, S. and Lemon, A. *Walking on Sticks*. La Mama Theatre, Melbourne, 12 October 1991.
Chadha, G. *Bend It Like Beckham*, Twentieth Century Fox, 2002.
Champagne, L., ed., *Out from Under: Texts by Women Performance Artists*, New York: Theatre Communications Group, 1990.
Champagne, L. 'Notes on autobiography and performance', *Women & Performance: A Journal of Feminist Theory* 19, 10: 1/2 (1999): 155–72.
Chi, J. and Kuckles. *Bran Nue Dae*, Sydney: Currency; Broome: Magabala Books, 1991.
Chodorow, N. 'The Enemy Outside: Thoughts on the Psychodynamnics of Extreme Violence With Special Attention to Men and Masculinity', in J. K. Gardiner, ed., *Masculinity Studies and Feminist Theory: New Directions*, New York: Columbia University Press, 2002, pp.235–60.
Churchill, C. *Churchill: Shorts*, London: Nick Hern Books, 1990.
Churchill, C. *Far Away*. London: Nick Hern Books, 2001.
Cixous, H. 'The Laugh of the Medusa' [1975], in Elaine Marks and Isabelle de Courtivron, eds, *New French Feminisms*, Brighton: Harvester Press, 1981, pp.245–64.
Cixous H. and Clément, C. *The Newly Born Woman*, trans. B. Wing, Manchester: Manchester University Press, 1987.
Comar, P. *The Human Body: Image and Emotion*, London: Thames & Hudson, 1999.
Connell, R. W. *Masculinities*, Berkeley: University of California Press, 1995.
Coote, A. and Campbell, B., *Sweet Freedom: The Struggle for Women's Liberation*, London: Picador, 1982.
Dahl, M. K. 'Postcolonial British Theatre: Black Voices at the Center', in J. E. Gainor, ed., *Imperialism and Theatre: Essays on World Theatre, Drama, and Performance*, London and New York: Routledge, 1995, pp.38–55.
De Lauretis, T. *Alice Doesn't: Feminism, Semiotics, Cinema*, Bloomington: Indiana University Press, 1984.
Desai, J., *Beyond Bollywood: The Cultural Politics of South Asian Diasporic Film*, London and New York: Routledge, 2004.
Diamond, E. "The Violence of 'We': Politicizing Identification", in J. G. Reinelt and J. R. Roach, eds, *Critical Theory and Performance*, Ann Arbor: The University of Michigan Press, 1992, pp.390–8.
Diamond, E. *Unmaking Mimesis*, London and New York: Routledge, 1997.
Dicker, R. and Piepmeier, A., eds, *Catching a Wave: Reclaiming Feminism for the 21st Century*, Boston, MA: North Eastern University Press, 2003.
Dolan, J. "Performance, utopia, and the 'utopian performative'", *Theatre Journal*, 53 (2001): 455–79.

Dovey, J. *FREAKSHOW: First Person Media and Factual Television*, London: Pluto Press, 2000.

Duden, B. *Disembodying Women: Perspectives on Pregnancy and the Unborn*, trans. L. Hoinacki, Cambridge, MA: Harvard University Press, 1993.

Edgar, D., ed., *State of Play*, London: Faber, 1999.

Elton, B. *Inconceivable*, New York: Bantam Doubleday Dell, new edn, 2000.

Elwes, C. 'Floating Femininity: A Look at Performance Art by Women', in S. Kent and J. Morreau, eds, *Women's Images of Men*, London: Writers and Readers Publishing, 1985, pp.164–93.

Evans, D. T. *Sexual Citizenship: The Material Construction of Sexualities*, London and New York: Routledge, 1993.

Faludi, S. *Backlash: The Undeclared War Against Women*, London: Vintage, 1992.

Faludi, S. *Stiffed: The Betrayal of Modern Man*, London: Vintage, 2000.

Fanger, I. 'Pulitzer Prize winner shakes off labels', *Christian Science Monitor*, 12 April 2002: 19.

Farquhar, D. '(M)Other Discourses', in G. Kirkup, L. Janes, K. Woodward, F. Hovenden, eds, *The Gendered Cyborg: A Reader*, London: Routledge, 1999, pp.209–20.

Feehily, S. *Duck*, London: Nick Hern Books, 2003.

Felski, R. *The Gender of Modernity*, Cambridge, MA, and London: Harvard University Press, 1995.

Fensham, R. 'Modernity and the /white Imaginary in Australian feminist theatre', *Hecate*, 29 (2003): 7–18.

Fensham, R. 'Making a mythopoetic theatre: Jenny Kemp as director of an imaginary future-past-present', *Australasian Drama Studies*, 44 (2004): 52–64.

Féral, J. 'Performance and theatricality: The subject demystified', *Modern Drama*, 25 (1983): 170–81.

Findlen, B. *Listen Up: Voices from the Next Feminist Generation*, 2nd edn, Seattle: Seal, 2001.

Finley, K. *The Constant State of Desire*, in L. Champagne, ed., *Out From Under: Texts by Women Performance Artists*, New York: Theatre Communications Group, 1990, pp.55–70.

Flax, J. *Thinking Fragments: Psychoanalysis, Feminism and Postmodernism in the Contemporary West*, Berkeley: University of California Press, 1990.

Forte, J. 'Women's performance art: Feminism and postmodernism', *Theatre Journal*, 40:2 (1988): 217–35.

Foucault, M. *The History of Sexuality, Volume 1, An Introduction*, trans. R. Hurley, London: Penguin Books, 1990.

Frankenberg, R. and Mani, L. 'Crosscurrents, Cross Talk: Race "Post-Coloniality" and the Politics of Location', in K.-K. Bhavnani, ed., *Feminism and Race*, Oxford: Oxford University Press, 2001, pp.479–516.

Frost, E. *Airsick*, London: Nick Hern Books, 2003.

Furse, A. *Your Essential Infertility Companion – a Users Guide to Tests, Technology and Treatment* [1997], London: Thorsons/Harper Collins, 2nd edn, 2001.

Furse, A. *Yerma's Eggs* (unpublished).

Gamble, S., ed., *The Routledge Critical Dictionary of Feminism and Postfeminism*, New York and London: Routledge, 2000.

Gammel, I., ed. *Confessional Politics: Women's Sexual Self-Representations in Life Writing and Popular Media*, Carbondale and Edwardsville: Southern Illinois University Press, 1999.

Gamman L. and Marshment, M., eds, *The Female Gaze: Women as Viewers of Popular Culture* [1988], London: The Women's Press, 1994.
George, K., ed., *Six Plays by Black and Asian Women Writers*, London: Aurora Metro Press, 1993.
Gillis. S., Howie, G. and Munford, R., eds, *Third Wave Feminism: A Critical Exploration*, London: Palgrave, 2004.
Gilroy, P., '"To Be Real": The Dissident Forms of Black Expressive Culture', in C. Ugwu, ed., *Let's Get It On: The Politics of Black Performance*, London: Institute of Contemporary Arts, 1995, pp.12–33.
Glaister, D. 'Welcome Queen Lear', The *Guardian*, 23 June 1997.
Gray, H. *Gray's Anatomy*, London: Bounty Books, 1977.
Green, N. 'Strange yet Stylish Headgear: Virtual Reality Consumption and the Construction of Gender', in E. Green and A. Adam, eds, *Virtual Gender: Technology, Consumption and Identity*, London: Routledge, 2001, pp.150–72.
Greer, G. *The Female Eunuch*, London: Flamingo, 2003.
Grehan, H. *Mapping Cultural Identity in Contemporary Australian Performance*, Brussels: Peter Lang, 2001.
Griffin, G. *Contemporary Black and Asian Women Playwrights in Britain*, Cambridge: Cambridge University Press, 2003.
Griffin, Kojo. 'Kojo Griffin: Work', http://www.millerblockgallery.com/artist_pages/Kojo_Griffin/Kojo_Griffin.html, accessed 3 May 2005.
Grosz, E. *Volatile Bodies: Towards a Corporeal Feminism*, Bloomington: Indiana University Press, 1994.
Hall, S. 'New Ethnicities', in H. A. Baker, Jr., M. Diawara and R. Lindenborg, eds, *Black Cultural Studies: A Reader*, Chicago and London: University of Chicago Press, 1996, pp.163–72.
Haraway, D. 'A Manifesto for Cyborgs', in L. J. Nicholson, ed., *Feminism/Postmodernism*, New York and London: Routledge, 1990, pp.190–233.
Haraway, D. *Simians, Cyborgs and Women: The Reinvention of Nature*, London: Free Association, 1991.
Haraway, D. 'The Virtual Speculum in the New World Order' in G. Kirkup, L. Janes, K. Woodward, F. Havenden, eds, *The Gendered Cyborg: A Reader*, London: Routledge, 1999, pp.221–245.
Harris, G. 'Introduction to Part Two', in L. Goodman with J. De Gay, eds, *The Routledge Reader In Gender And Performance*, London and New York: Routledge, 1998, pp.55–9.
Harris, G. *Staging Femininities: Performance and Performativity*, Manchester: Manchester University Press, 1999.
Harris, G. 'Foreshadowings and after blows: Fiona Templeton's *The Medead* (in progress)', *Studies in Theatre and performance*, 23: 3 (2004): 165–77.
Harwood, Kate, ed., *First Run*, London: Nick Hern Books, 1989.
Haynes, R. *Seeking the Centre: The Australian Desert in Literature, Art and Film*, Sydney: Cambridge University Press, 1998.
Heidegger, M. *The Question Concerning Technology and other Essays*, New York and London: Harper Row 1977.
Hershman-Leeson, L., ed., *Clicking In: Hot Links to a Digital Culture*, Seattle: Bay Press, 1996.
Hetherington, K. *Expressions of Identity: Space, Performance, Politics*, London: Sage, 1998.
Heywood, L. and Drake, J., eds, *Third Wave Agenda: Being Feminist, Doing Feminism*, Minneapolis: University of Minnesota Press, 1997.

Hobsbawm, E. J. 'Ethnicity and Nationalism in Europe Today', in G. Balakrishnan, ed., *Mapping the Nation*, London and New York: Verso, 1996, pp.255–66.
Holland, P. *English Shakespeares: Shakespeare On The English Stage In The 1990s*, Cambridge, New York, Melbourne: Cambridge University Press, 1997.
hooks, b. *Feminist Theory: From Margin to Center*, Boston, MA: South End Press, 1984.
hooks, b. *Talking Back: Thinking Feminist, Thinking Black*, Boston, MA: South End Press, 1989.
Hornby, R. *The Hudson Review* (Winter 1996): 641–4.
Howard, J., Rackin, P. *Engendering A Nation: A Feminist Account Of Shakespeare's English Histories*, Routledge: London and New York, 1997.
Hughes, H. and Román, D., eds, *O Solo Homo: The New Queer Performance*, New York: Grove Press, 1998.
Hughes, S. 'Slam poets – Information on United Kingdom performance Poets', http://www.flippedeye.net/slampoets/ukpoets.html, accessed 3 May 2005.
Huxley, A. *Brave New World* [1934], London: Harper Collins, 1994.
Ibsen, H. *Hedda Gabler* trans. C. Hampton, London: Samuel French, 1989.
Isaak, J. *Feminism and Contemporary Art: The Revolutionary Power of Women's Laughter*, London and New York: Routledge, 1996.
James S. M., ed., *Theorising Black Feminisms: The Visionary Pragmatism of Black Women*, London and New York: Routledge, 1993.
Janaczewska, N. *Historia*, in P. Tait and E. Schafer, eds, *Australian Women's Drama: Texts and Feminisms*, Sydney: Currency, 1997, pp.254–84.
Jarman, N. *Material Conflicts: Parades And Visual Displays In Northern Ireland*, Oxford and New York: Berg, 1997.
Jelinek, E. *Krankheit oder Moderne Frauen*, Köln: Prometh Verlag, 1987.
Jones, J. L. 'Sista Docta', in L. C. Miller, J. Taylor and M. H. Carver, eds, *Voices Made Flesh: Performing Women's Autobiography*, Wisconsin: University of Wisconsin Press, 2003, pp.237–57.
Joseph, M. 'Borders Outside the State: Black British Women and the Limits of Citizenship', in P. Phelan and J. Lane, eds, *Ends of Performance*, New York: New York University Press, 1998, pp.197–213.
Kalb, J. *Theater*, 31:3 (2000): 14.
Kane, S. *4.48 Psychosis*, in *Sarah Kane: Complete Plays*, London: Methuen Drama, 2001, pp.203–45.
Kane, S. *Sarah Kane: Complete Plays*, London: Methuen Drama, 2001.
Kellaway, K. 'Good for a girl', The *Observer*, London, 31 October 2004.
Kemp, J. *Still Angela*, Sydney: Currency, 2002.
Kemp, J. 'A Dialogue with Disjunction', in Virginia Baxter, ed., *Telling Time: Celebrating Ten Years of Women Writing for Performance*, Sydney: Playworks, 1996, pp.28–31.
Khan-Din, A. *East Is East*, London: Nick Hern Books, 1997.
Kirkup, G., Janes L., Woodward, K. and Hovenden, F., eds, *The Gendered Cyborg: A Reader*, London and New York: Routledge and Open University Press, 2000.
Koenig, R. 'Theatre Reviews', *Independent*, 5 June 1995.
Kolodny, A. *The Lay of the Land: Metaphor as Experience and History in American Life and Letters*, Chapel Hill: University of North Carolina Press, 1975.
Kozel, S. 'Multi Medea: Feminist Performance Using Multimedia Technology', in L. Goodman and J. De Gay, eds, *The Routledge Reader in Gender and Performance*, London and New York: Routledge, 1998, pp.299–302.

Kron, L. *2.5 Minute Ride and 101 Humiliating Stories*, New York: Theatre Communications Group, 2001.
Kron, L. *Well*, unpublished script (courtesy of Kron).
Kwei-Armah, K. Interview, *Today*, broadcast BBC Radio 4, 13 December 2004.
Kwei-Armah, K. Interview with director Angus Jackson, in programme notes, *Elmina's Kitchen*, regional tour United Kingdom, 2005.
Law, D. and Peaty S. *Angels and Mechanics*, Catalogue, London Arts Board/Arts Council, 1996.
Lawford, J. N., Chaplin, A. and Archer, R. *Ningali: The Story So Far*, unpublished ms, 1994.
Lefebvre, H. *The Production of Space*, Oxford: Blackwell, 1991.
Lemon, A. and Cathcart, S. *Tiger Country*, unpublished ts, 1996.
Lochhead, L. *Perfect Days*, London: Nick Hern Books, 1999.
Lorca, F. G. *Yerma, Plays One*, trans. P. Luke, London: Methuen, 2000.
Lyndon, N. *No More Sex War: The Failures of Feminism*, London: Sinclair-Stevenson, 1992.
McRobbie, A. *Feminism and Youth Culture* [1991] 2nd edn, Basingstoke: Macmillan – now Palgrave Macmillan, 2000.
McRobbie, A. *In the Culture Society: Art, Fashion and Popular Music*, London: Routledge,1999.
McRobbie, A. and McCabe, T., eds, *Feminism for Girls: An Adventure Story* London: Routledge, 1981.
Mahone, S., ed., *Moon Marked and Touched By Sun: Plays By African American Women*, New York: Theatre Communications Group, 1994.
Mama, A., 'Black Women and the British State: Race, Class and Gender Analysis for the 1990s', in P. Braham, A. Rattansi and R. Skellington, eds, *Racism and Anti-Racism: Inequalities, Opportunities and Policies*, London: Sage Publications, 1992, pp.79–101.
Mangan, J. 'Many Faces, and All of them Angela', *The Age*, 9 April 2002: Culture 4.
Mangan, M. *Staging Masculinities*, Basingstoke: Palgrave Macmillan, 2003.
Martin, E. *The Woman in The Body: A Cultural Analysis of Reproduction*, Boston, MA: Beacon Press, 1992.
Massey, D. *Space, Place and Gender*, Cambridge: Polity, 1994.
Medhurst A. 'Camp', in A. Medhurst and S. R. Munt, eds, *Lesbian and Gay Studies: A Critical Introduction*, London: Cassells, 1997, pp.274–93.
Mercer, K. *Welcome to the Jungle: New Positions in Black Cultural Studies*, New York and London: Routledge, 1994.
Merleau-Ponty, M. *Phenomenology of Perception* trans. C. Smith, London, New York: Routledge, 2002.
Meyer, R. 'Have you heard the one about the lesbian who goes to the Supreme Court: Holly Hughes and the case against censorship', *Theatre Journal*, 52 (2000): 543–52.
Miller, A. *The Crucible, A Play In Four Acts*, Middlesex, New York, Victoria, Ontario, Auckland: Penguin Books, 1982.
Miller, D. A. *Bringing Out Roland Barthes*, Berkeley: University of California Press, 1992.
Miller, L. C. Taylor, J. and Carver, M. H., eds, *Voices Made Flesh: Performing Women's Autobiography*, Wisconsin: University of Wisconsin Press, 2003.
Monk, C. 'Men in the 90s', in R. Murphy, ed., *British Cinema of the 90s*, London: *BFI*, 2000, pp.156–66.

Muñoz, J. 'The Future in the Present: Sexual Avant-Gardes and the Performance of Utopia', in D. Pease and R. Weigman, eds, *The Future of American Studies*, Durham, NC, and London: Duke University Press, 2002, pp.77–93.

New York State Council for the Arts. 'Report on the Status of Women: A Limited Engagement?',http://www.americantheaterweb.com/nysca/opening.htm, viewed 26 April 2004.

Nilsson, L. and Hamberger, L. *A Child Is Born*, London: Doubleday, 1990.

Pagels, E. *The Gnostic Gospels*, London: Penguin Books, 1990.

Paglia, C. *Sexual Personae: Art and Decadence from Nefertiti to Emily Dickinson*, New Haven: Yale University Press, 1990.

Parker, A. et al., eds, *Nationalisms and Sexualities*, New York: Routledge, 1992.

Park-Fuller, L. 'A Clean Breast of It', in L. C. Miller, J. Taylor and M. H. Carver, eds, *Voices Made Flesh: Performing Women's Autobiography*, Wisconsin: University of Wisconsin Press, 2003, pp.215–36.

Parks, S. L. *The America Play and Other Works*, New York: Theatre Communications Group, 1995.

Parmar, P. 'Other Kinds of Dreams' [1989], in H. S. Mirza, ed., *Black British Feminism: A Reader*, London and New York: Routledge, 1997, pp.67–9.

Pellegrini, A. 'Staging Sexual Injury: How I Learned to Drive', in J. Reinelt and J. Roach, eds, *Contemporary Theory and Performance*, Ann Arbor: University of Michigan Press, 2006.

Phelan, P. 'The Ontology of Performance: Representation without Reproduction', in *Unmarked: The Politics of Performance*, London and New York: Routledge, 1993, pp.146–66.

Pinnock, W., *Talking in Tongues*, in Y. Brewster, ed., *Black Plays: Three*, London: Methuen, 1987, pp.171–227.

Plummer, K. *Telling Sexual Stories: Power, Change and Social Worlds*, London: Routledge, 1995.

Pollack P. R. 'Foetal Images: The Power of Visual Culture in the Politics of Reproduction', in G. Kirkup, L. Janes, K. Woodward, F. Havenden, eds, *The Gendered Cyborg: A Reader*, London: Routledge, 2000, pp.171–92.

Politt, K. and Baumgardner, J. 'Afterword: A Correspondence between Katha Politt and Jennifer Baumgardner', in R. Dicker and A. Piepmeier, eds, *Catching a Wave: Reclaiming Feminism for the 21st Century*, Boston, MA: North Eastern University Press, pp.309–19.

Ponnuswami, M. 'Small Island People: Black British Women and the Performance of Retrieval', in E. Aston and J. Reinelt, eds, *Cambridge Companion to Modern British Women Playwrights*, Cambridge: Cambridge University Press, 2000, pp.217–34.

Rabkin, G. 'Is There a Text on This Stage? Theatre, Authorship, Interpretation', in R. Schneider and G. Cody, eds, *Re:Direction: A Theoretical And Practical Guide*, London and New York: Routledge/TDR (2002), pp.319–32.

Reckitt, H., ed., *Art and Feminism*, London: Phaidon Press, 2001.

Reinelt, J., 'Staging the invisible: The crisis of visibility in theatrical representation', *Text and Performance Quarterly*, 14: 2, (1994): 97–107.

Reno. 'Reno: Rebel Without a Pause: Unrestrained Reflections on September 11th'. <http://www.citizenreno.com/index_rebel.html>. Viewed 6 Dec. 2004.

Rich, A. *Of Woman Born: Motherhood as Experience and Institution*, London: Virago, 1977.

Roach, J. *Cities of the Dead: Circum-Atlantic Performance*, New York: Columbia University Press, 1996.
Roach, J. 'Deep Skin: Reconstructing Congo Square', in H. J. Elam, Jr. and D. Krasern, eds, *African American Performance and Theater History: A Critical Reader*, Oxford and New York: Oxford University Press, 2001, pp.101–13.
Rogoff, I. *Terra Infirma: Geography's Visual Culture*, London: Routledge, 2000.
Roiphe, K. *The Morning After: Sex, Fear, and Feminism on College Campus*, Bolton, MA: Little, Brown, 1993.
Rose, G. 'Some Notes Towards Thinking about the Spaces of the Future', in Jon Bird, Barry Curtis, Tim Putnam, George Robertson, Lisa Tickner, eds, *Mapping the Futures: Local Cultures, Global Change*, London: Routledge, 1993, pp.70–83.
Roth, M. 'Autobiography, Theater, Mysticism and Politics: Women's Performance Art in Southern California', in C. E. Loeffler, ed., *Performance Anthology: Source Book of California Performance Art*, San Francisco: Contemporary Arts Press, 1980, pp.463–89.
Roth, M., ed., *Rachel Rosenthal*, Baltimore: The Johns Hopkins University Press, 1997.
Rowbotham, S. *The Past is Before Us: Feminism in Action Since the 1960s*, London: Pandora Press, 1989.
Rowbotham, S., Segal, L. and Wainwright. H., eds, *Beyond the Fragments: Feminism and the Making of Socialism*, London: The Merlin Press, 1979.
Rowe-Finkbeiner, K. *The F Word; Feminism in Jeopardy*, Emeryville, CA: Seal Press, 2004.
Russell, M., ed., *Out of Character: Rants, Raves, and Monologues from Today's Top Performance Artists*, London: Bantam Books, 1997.
Rutherford, J. *The Gauche Intruder: Freud, Lacan, and the White Australian Fantasy*, Melbourne: Melbourne University Press, 2000.
Rutter, C. 'Fiona Shaw's Richard II: The Girl As Player-King As Comic', *Shakespeare Quarterly*, 48 (1997): 314–24.
Said, E. *Culture and Imperialism*, New York: Knopf, 1993.
Saunders, G. *'Love Me or Kill Me': Sarah Kane and the Theatre of Extremes*, Manchester: Manchester University Press, 2002.
Savran, D. 'The Wooster Group, Arthur Miller And *The Crucible*', *The Drama Review*, 29 (1985): 99–109.
Savran, D. *Breaking The Rules, The Wooster Group*, New York: Theatre Communications Group, 1988.
Savran, D. *A Queer Sort of Materialism: Recontextualizing American Theater*, Ann Arbor: University of Michigan, 2003.
Schaffer, K. *Women and the Bush: Forces of Desire in the Australian Cultural Tradition*, Cambridge: Cambridge University Press, 1988.
Schawtz, H. *The Culture of The Copy: Striking Likenesses and Unreasonable Facsimiles*, Cambridge, MA: MIT Press, 1996, pp.261–73.
Schneider, R. and Cody G., eds, *Re:Direction: A Theoretical And Practical Guide*, London and New York: Routledge/TDR, 2002.
Schulman, S. *My American History: Lesbian and Gay Life During the Reagan/Bush Years*, New York: Routledge, 1994.
Scott, J. W. 'Experience', in J. Butler and J. W. Scott, eds, *Feminists Theorize the Political*, London: Routledge, 1992, pp.22–41.
Scott-Norman, F. 'Review of *Black Sequin Dress* by Jenny Kemp', *Bulletin*, 9 April 1996. Reprinted in *ANZTR* March 1996: 52.

Sedgwick, E. K. *Touching Feeling: Affect, Pedagogy, Performativity*, Durham, NC, and London: Duke University Press, 2003.

Segal, L., Rowbotham, S. and Wainwright, H. *Beyond the Fragments: Feminism and the Making of Socialism*, London: The Merlin Press, 1979.

Shakespeare, W. *The Tragedy Of King Richard II*, London, New York, Victoria Toronto, New Delhi, Auckland, Johannesburg: Penguin Books, 1997.

Siegel, D. L. 'Reading Between the Waves', in L. Heywood and J. Drake, eds, *Third Wave Agenda: Being Feminist, Doing Feminism*, Minneapolis: University of Minnesota Press, 1997, pp.40–54.

Silver, L., M. *Remaking Eden: Cloning, Genetic Engineering and the Future of Humankind?*, New York: Avon Books, 1997.

Sinfield, A. 'Diaspora and Hybridity: Queer Identities and the Ethnicity Model', in N. Mirzoeff, ed., *Diaspora And Visual Culture: Representing Africans and Jews*, London and New York: Routledge, 2000, pp.95–114.

Smith, S. 'The Autobiographical Manifesto: Identities, Temporalities, Politics', in S. Neuman, ed., *Autobiography and Questions of Gender*, London: Frank Cass, 1991, pp.186–212.

Smith, S. 'Autobiographical Manifestos' [1993], in S. Smith and J. Watson, eds, *Women, Autobiography, Theory: A Reader*, Wisconsin: University of Wisconsin Press, 1998, pp.433–40.

Solomon, A. *Redressing The Canon: Essays On Theater And Gender*, New York and London: Routledge, 1997.

Spry, T. "Illustrated Woman: Autoperformance in 'Skins: A Daughter's (Re)Construction of Cancer' and 'Tattoo Stories: A Postscript to "Skins"', in L. C. Miller, J. Taylor and M. H. Carver, eds, *Voices Made Flesh: Performing Women's Autobiography*, Wisconsin: Wisconsin University Press, 2003, pp.157–91.

Stanley, L. *The Auto/Biographical I: The Theory and Practice of Feminist Auto/Biography*, Manchester: Manchester University Press, 1992.

States, B. *Great Reckonings In Little Rooms: On The Phenomenology Of Theater*, Berkeley, Los Angeles, London: University of California Press, 1985.

Steinberg, D. L. 'Feminist Approaches to Science, Medicine and Technology', in G. Kirkup, L. Janes, K. Woodward, F. Havenden, eds, *The Gendered Cyborg: A Reader*, London: Routledge, 2000, pp.193–208.

Stowell, S. 'Rehabilitating realism', *Journal of Dramatic Theory and Criticism*, 6: 2 (Spring 1992): 81–8.

SuAndi. *The Story of M*, in SuAndi, ed., *4 for More*, Manchester: Black Arts Alliance, 2002, pp.1–18.

Syal, M. *Bhaji on the Beach*, dir. Gurinder Chadha, First Look, 1993.

Syal, M. *My Sister-Wife*, in K. George, ed., *Six Plays by Black and Asian Women Writers*, London: Aurora Metro Press, 1993, pp.111–58.

Taylor, D. *Archive and the Repertoire: Performing Cultural Memory in the Americas*, Durham, NC: Duke University Press, 2003.

Thomas, D. *Not Guilty: In Defence of Modern Man*, London: Weidenfeld & Nicolson,1993.

Townsend, S. *Ten Tiny Fingers, Nine Tiny Toes*, London: Methuen Drama, 1990.

Troyano, A. *I, Carmelita Tropicana: Performing Between Cultures*, Boston, MA: Beacon Press, 2000.

'Urban Griot', http://www.blackauthors.com/, accessed 3 May 2005.

Viner, K. 'The Personal Is Still Political', in N. Walter, ed., *On the Move: Feminism for a New Generation*, London: Virago Press, 2000, pp.10–26.

Vogel, P. *How I Learned to Drive*, in *The Mammary Plays*. New York: Theatre Communications Group, 1998.

Walker, R., ed., *To Be Real: Telling the Truth and Changing the Face of Feminism*, New York: Anchor Books, 1995.

Wajcman, J. *Techno Feminism*, Cambridge: Polity Press, 2004.

Walter, N. *The New Feminism*, London: Virago, 1998.

Wandor, M. *Wanted*, London: Playbooks, 1988.

Wandor, M. *Aid Thy Neighbour*, in M. Wandor, *Five Plays*, London: Journeyman Press, 1984.

Wardle, I. 'The woman who would be king', *Independent On Sunday*, 11 June 1995.

Wertenbaker, T. *The Break of Day*, London: Faber, 1995.

Whelehan, I. *Modern Feminist Thought: From the Second Wave to 'Post-Feminism'*, Edinburgh: Edinburgh University Press, 1995.

Whelehan, I. *Overloaded: Popular Culture and the Future of Feminism*, London: The Women's Press, 2000.

Wilding, F. 'The Feminist Art Programes at Fresno and Calarts, 1970–75', in N. Broude and M. D. Garrard, eds, *The Power of Feminist Art: The American Movement of the 1970s, History and Impact*, New York: Harry N. Abrams, 1994, pp.32–47.

Wittgenstein, L. *Culture and Value*, G. H. Von Wright, ed., trans. P. Winch, Cambridge, MA, and Oxford: Blackwell, 1980.

Wolf, N. *The Beauty Myth: How Images of Beauty are Used Against Women*, London: Vintage, 1991.

Wolf, N. *Fire with Fire: The New Female Power and How it Will Change the Twenty First Century*, London: Vintage, 1994.

Wolverton, T. *Insurgent Muse: Life and Art at the Woman's Building*, San Francisco: City Lights Books, 2002.

Wong, K. S. 'Pranks and Fake Porn: Doing Feminism My Way', in R. Dicker and A. Piepmeier, eds, *Catching a Wave: Reclaiming Feminism for the 21st Century*, Boston, MA: North Western University Press, 2003, pp.294–307.

Worthen, W. B. *Shakespeare And The Authority Of Performance*, Cambridge, New York, Melbourne: Cambridge University Press, 1997.

Yuval-Davis, N. 'Women, citizenship, and difference', *Feminist Review*, 57 (Autumn 1997): 4–27.

Zahir, S., 'Goodness Gracious Me', in A. Donnell, ed., *Companion to Contemporary Black British Culture*, London and New York: Routledge, 2002, pp.128.

Zimmerman, B. 'The Politics of Transliteration: Lesbian Personal Narratives', in. E. B. Freedman, B. C. Gap, S. L. Johnson, K. M. Weston, eds, *The Lesbian Issue – Essays from Signs*, Chicago: University of Chicago Press, 1985, pp.251–70.

Index